Why Science?

Why Science?

R. STEPHEN WHITE

NOVA SCIENCE PUBLISHERS, INC.
Commack, NY

Assistant Vice President/Art Director: Maria Ester Hawrys
Office Manager: Annette Hellinger
Graphics: Frank Grucci
Acquisitions Editor: Tatiana Shohov
Book Production: Ludmila Kwartiroff, Christine Mathosian, Maria A. Olmsted and Tammy Sauter
Editorial Production: Susan Boriotti
Circulation: Cathy DeGregory and Maryanne Schmidt

Library of Congress Cataloging-in-Publication Data

White, R. Stephen, 1920-
Why science / by R. Stephen White
p. cm.
Includes index.
ISBN 1-56072-531-1
1. Science—Social aspects. 2. Science—Popular works. I. Title.
Q175.5.W48 1997 97-39032
500—dc21 CIP

6080 Jericho Turnpike, Suite 207
Commack, New York 11725
Tele. 516-499-3103 Fax 516-499-3146
E-Mail: Novascience@earthlink.net
Web Site: http://www.nexusworld.com/nova

Printed in the United States of America

CONTENTS

Introduction

The state of science and technology perceived by the activists seems widely divergent from mine, and I think from that held by most other scientists. But the 'nayactivists' and 'prophets of doom' have been controlling the agenda. It's no wonder the public is confused and concerned, afraid of the future--too timid to demand evidence and too paralyzed to fight back.

The public gets its 'science' from TV talk shows--from programs like "The UFO Cover Up? - Area 51" on 'Larry King Live' or one of his 'Special Science' programs broadcast from the 'extraterrestrial highway' at the scholarly research center of Rachel, Nevada. I'm always surprised that the audiences still love entertainment that gives them the same stories by the same 'UFO experts' with never a shred of evidence, only talk. No alien spacecraft, no aliens, not even a clear photo of alien abduction. The experts' excuse: we can't give you the evidence, it's suppressed by a 'conspiracy', a vast government cover up.

Reporters from newspapers are good at reporting new discoveries in science. Using press releases from authors at times of publication, and interviews to clarify questions--reporters write interesting and informative stories of value to readers.

On matters of controversy, their record is spotty at best. Too often normal mainstream science is not considered newsworthy. 'News' is maverick scientists (or non-scientists) challenging the establishment. So the reporter interviews two on the pro and two on the con and reports what they say, too often emphasizing the sensational. It matters not that the mavericks' views are supported by poor evidence or no evidence at all and are held by only a small fraction of the knowledgeable scientists. The reader gets a grossly distorted picture of the issue. He is confused--tends to believe iconoclasts' views that crises abound--although none really exist.

Nor does the TV viewer or newspaper reader or the public know where to go for help. On matters of politics they can listen to TV political interview programs, read the newspaper editorials and letters to the editor to get different views. But on TV, except for a few programs like 'NOVA' on a limited number of subjects, they find entertainment, not science. Newspaper columns can be found on medicine, cooking, gardening and all sorts of things, even astrology, but rarely on science--at least on the kind that will help make intelligent decisions on issues involving science and technology.

For the last two years, I've been writing the column 'Science Views' for the Santa Barbara News Press. This is my attempt to furnish information that the public may find useful in making difficult decisions. I am well aware that these views are not shared by all people or even all scientists. I'm constantly reminded of this by letters to the editor, personal letters and calls.

"Why Science?" is written for the millions of people who want evidence for their views. Much evidence may be found in the science columns that make up the book. The columns are my views which, I think, in most cases are also the views of conventional science. The issues involving science in Santa Barbara, California, the nation and even the rest of the world are similar. I hope the columns in the book will supply some of the arguments sorely needed by the public.

PUBLIC WORRIES

The Public is told by the 'population activists' that the world's population is already out of control. We agree that an ever-increasing population, indeed, is a concern and must be controlled. But, in fact, the increases in population growth are decreasing all over the earth, even in most of the underdeveloped nations--while the birth rates in many of the most developed nations have dropped below a sustainable population. Most experts now expect the world's population to reach a constant level by the year 2050.

The 'energy activists' tell us we must immediately switch from energies using coal, oil, gas and nuclear to the renewable energies using solar collectors, photovoltaics, windmills, geothermal and biomass. However, there are serious problems with each of the renewables that are too often ignored by the enthusiasts. Each has produced less than one percent of the world's required power. And that is not likely to change in the foreseeable future.

Electric battery car 'enthusiasts' would have us switch from gasoline to battery power. The California Air Resources Board in 1990 mandated that 2% of all autos for sale in 1998 be exhaust-pollution-free (electric battery powered). This was to be increased to 10% by the year 2003. But they repealed the mandate in March, 1996, except for the 2003 date. It was clear that no batteries would be available that would satisfy the requirements of distance of travel, speed, acceleration, safety and convenience at a price that the public would pay. No person, organization or government agency can mandate a new science or technology break-through. They come only when the science is known and the technology is ready.

Agriculture in America has been one of the most successful ventures in the history of the human race. Less than two percent of the U.S. population supplies 10 percent of the world's wheat, 40 percent of its corn and 45 percent of its soy beans. The United States ships abroad 30 percent of the world's exports of wheat and 60 percent of its corn and soy beans. The United States not only feeds its own population but a significant fraction of the rest of the world. This is possible because of the many scientific and technical innovations introduced on the farm.

Mechanization of farms occurred mostly after World War I. Tractors and trucks replaced horses and combines replaced reapers. New plant hybrids, methods of plowing and cultivation, irrigation, fertilizers and insecticides increased yields by factors of 10 and decreased man hours per bushel by much more.

Yet the 'organic food activists' would have us return to the out-dated practices of the past. No 'chemical' fertilizers, only manure--although, of course, all fertilizers, including manure, are made up of chemical compounds. And there are not enough animals on earth to produce the replacement manure. Insecticides and fungicides will be replaced by natural enemies, most yet to be identified and tested.

We all agree it is necessary to protect people and animals against dangerous pesticides and preserve the environment. The Environmental Protection Agency, in the past, used the sensible 'negligible risk' that permitted small, harmless amounts of pesticides in processed food and other products. However, the Clinton administration is carrying safety to an unacceptable extreme as it considers prohibiting 'any detectable amount' in processed foods.

To determine whether a chemical is a carcinogen, usually massive doses are fed to rodents. Such large doses can destroy the animal's immune system, so the results have little value for determining the effects of very low doses consumed by humans. Also the procedure of extrapolation from very high doses in rats to microscopic doses in humans is challenged by many biologists.

A recent study from the National Research Council indicates that the level of naturally occurring carcinogens in unprocessed food is higher than the level of added synthetic carcinogens. And the natural ones are unregulated. The report concludes that the natural and synthetic chemicals in the diet are at such low levels they are unlikely to pose an appreciable cancer risk.

The 'animal activists' disobey the law as they disrupt animal experiments designed to save human life. Insulin for diabetics was developed with research on dogs and a vaccine for polio was found with research on monkeys. Drugs for cancer, AIDS, Parkinson's disease and many others are dependent on experiments with mice and other rodents.

The 'alternative medicine activists' would have the medical profession accept treatments from practitioners with little medical training and take drugs without scientific tests. This, although the longevity of Americans has increased nearly 50 percent in my lifetime, is continuing to rise and the United States is the world's leader in medical research.

The Endangered Species Act was passed by Congress in 1973 to protect the species' habitats. The Act is again under review by Congress. It's obvious that the species are best protected when their habitats are of maximum size. So the 'wildlife activists' are vigorously lobbying to keep the Act in its present form. The Act might be workable if the public's only goal were to protect endangered species and their habitats; and if Fish and Wildlife Service and other state and local organizations were properly staffed and funded; and private property owners were properly compensated for forced participation. Because none of these conditions has been met, a fair law is needed that protects citizen rights as well as the habitats of endangered species.

A recent attack on science has come from an unexpected quarter, from the 'academic left', a loose amalgam of postmodernism, feminism, radical environmentalism and multiculturalism. They claim that scientific knowledge is not a version of universal truth that is the same at all times and places and that there are other ways of knowing. They incorrectly interpret the 'uncertainty principle' and 'chaos' in physics and mathematics as discrediting physics predictability.

The 'social constructionists' claim that science and scientific methods are only social constructs and that truth is subjective. Yet not one would defy the laws of physics by jumping from the roof of a three story building or drive a car into a concrete wall at 70 mph.

SOMETIMES FORGOTTEN ADVANCES IN SCIENCE AND TECHNOLOGY

First a reminder of the technology and science of my youth in the late 1920s and 30s and the advances for comparison with those of today.

Typical transportation of the 20s was a model T Ford with celluloid and cloth snap-on windows, clutch for reverse and spark adjustment and gas feed on the steering wheel column. Most roads were dirt that developed deep ruts after rains and wash board surfaces when dry. The model T coupe was top heavy and dangerous on sharp corners and when driven over 30 mph on the roads. The cars had no air conditioning, radios, stereos, tape or disc drives; no seat belts, air bags or comfortable seats. Arm signals were required and the windshield wiper was rotated by hand. The Ford V-8 of the 1930s was a major improvement.

Most people traveled cross country by train. The few airplanes carrying freight and passengers were propeller driven and airmail was more expensive than mail sent on the ground. My first travel by commercial plane, propeller, was in the late 1940s. Jet propulsion was developed during World War II.

My father, at various times in his career, was a public school teacher, principal and superintendent, in a number of small Kansas towns. In the summers and a few winters we lived on a farm. A windmill pumped water into tanks for the stock and the house. Like many country homes of that time, it lacked inside plumbing. A small wind-charger furnished power for a radio and dim lights. We had a crank telephone with a party line; cooked on an iron range that burned wood and coal; and that, with a pot-bellied iron stove, furnished the heat.

In the early 1930s, our ice-box was replaced by a refrigerator fueled by butane. For entertainment record players were available but not yet TV. Life improved dramatically in the late 1930s when the Rural Electrification Agency brought electricity to the farm. The small towns where we lived had most of the utilities available today. Sound and Technicolor at the movies were still quite new.

A. Explosion of the Information Sciences--College majors in chemistry, physics and engineering in the 30s and 40s were easily identified by slide rules attached to their belts. Primitive mechanical calculators powered by hand were available for simple calculations. These were replaced in the late 40s by electrical calculators that still took seconds to evaluate square roots.

The first digital computer, built in 1946, filled a large room with thousands of vacuum tubes. The invention of transistors by solid state physicists permitted great reduction in size. Integrated circuits followed. Minicomputers were available in the 1960s followed by microprocessors in the 1970s. Computers today handle information at speeds greater than a billion bytes a second and maximum limits have yet to be reached.

A whopping 36 percent of the households in the United States now have personal computers. In southern California it is even higher where 46 percent of the families in Los Angeles and Orange counties own PCs. More than 60 percent of those polled believe the new information technology is leading to a better life. Users of Internet and the world-wide-web are increasing rapidly. Integrated home entertainment centers will soon be commonplace. Computers have been customary at the check-out counters for years. The computer is invaluable for writers and research in all disciplines. Computer graphics have largely replaced the draftsmen and the technical artists. Scientific instruments and industrial processes are run by computer. Large companies are positioning for control of communications, and entertainment world-wide. We are living in the computer age.

B. Discoveries in Nuclear Physics--Experiments carried out by physicists in the early 1900s showed that atoms were not the smallest pieces of matter after all. With the discovery of the neutron in 1932 it was clear that atoms were made up of protons and neutrons in the core called the nucleus. Electrons circled the nucleus much like planets around the sun. We now know that protons and neutrons are composed of even smaller units called quarks.

The new science of nuclear physics benefits human life in many ways. In the United States nuclear medicine treats over ten million patients per year, one-third of the total number hospitalized.

In 1938, the fission reaction was discovered. A uranium 235 nucleus can capture a neutron and split into two or more smaller nuclei and emit a large amount of energy, a million times that obtained from reactions with atoms. It was immediately apparent that this energy could be used in controlled reactions, just as in controlled chemical reactions using coal, oil or gas to generate electricity; and in run-away reactions to produce gigantic explosions similar to dynamite in run-away chemical reactions.

Correctly used, nuclear reactor-produced energy could be the energy of the future. It is cheaper, cleaner and safer for generating electricity than any of the other possible options. However, the 'anti-nuclear activists' have done their job well. Because of the fears of the public, no new reactors have been authorized in the United States in the last 20 years. This is in spite of the fact that 109 US commercial reactors have 1,200 operating years of experience without a single death caused by accidents. The critics point to the Three Mile Island nuclear reactor accident as proof that the reactors are unsafe. But it proved the contrary. Not one person was killed or injured by the accident.

The opposition by 'anti-nuclear activists' to both low and high-level radioactive waste storage facilities is aimed at stopping the use of nuclear energy. That is not in the public interest. The low-level wastes from hospitals and medical labs and from university medical research are currently stored on the local sites. Almost everyone agrees that permanent facilities should replace the temporary ones. Yet every suggested permanent site is vetoed by some single-issue activist group. For example, the Ward Valley site in California is hung up by the refusal of Secretary of Interior, Bruce Babbitt, to transfer the federal land to California. Senator Barbara Boxer, outspoken critic of Ward Valley, apparently considers storage at UC San Francisco, in the middle of the city, of less danger to the public than the barren area in the Mojave Desert.

The difficulty in finding a repository for high level waste is legend. The Nuclear Waste Policy Act of 1982 gave the Department of Energy the responsibility for selection and development of a site. The choice was Yucca Mountain in the Nevada Test Site near Death Valley. After years of study and mountains of reports the site still has not been authorized. There is no question that high level wastes should be moved away from the reactor sites where they are currently stored. The Yucca Mountain site will contain the used reactor fuel rods safely for the next 10,000 years. However, indecision and inaction have paralyzed our nation. The 'antinuclear activists' have succeeded in stalling the major 20th century scientific contribution to the welfare of the world. Only by vigorous education of the public in science can we hope to reverse this trend.

C. Progress in Astronomy and Astrophysics--In my lifetime the increase in knowledge of astronomy and astrophysics has been phenomenal. The same is true for several of the other sciences. It is hard for us to realize that it wasn't until the 1920s that astronomers agreed that there were other galaxies in the universe than our own. Only optical telescopes were used for astronomical observations before World War II.

Developments in radar and other radio-wave techniques during the war led to the invention of radio-wave telescopes for looking at the stars and other galactic and extragalactic objects by the radio waves they emitted. This was followed by telescopes invented for observing radiation from celestial objects in all parts of the electromagnetic spectrum. These telescopes were carried aloft as required on high flying balloons, rockets and satellites.

Evidence is now very strong for the Big Bang origin of the universe. The Hubble expansion of the universe, microwave cosmic background radiation and the abundance of the isotopes of hydrogen and the other light elements all support the Big Bang origin that occurred about 15 billion years ago. The formation of galaxies and clusters of galaxies are under study with observations from the Hubble space telescope and the giant Keck and other large telescopes on the ground. Discoveries of pulsars, neutron stars, black holes, quasars and other strange objects are under intense observational and theoretical study.

Many consider the space program, crowned by the Apollo 11 landing of astronauts on the moon in 1969, the greatest technological achievement of the 20th century. The production of the V-2 rockets by Germany during World War II, led to the development of giant boosters by the United States and the

USSR. The successes of the earth-orbiting satellites--Sputnik by the USSR in 1957 and Explorer by the United States in 1958--enabled President Kennedy, in 1961, to commit America to the landing of astronauts on the moon.

The new field of 'space physics' was opened for exploration. The radiation belts around the earth were discovered and the particles identified. Unmanned space probes with a variety of detectors were sent to the planets to measure their properties. More recently large unmanned satellite observatories for optical, infrared, ultraviolet, x-ray and gamma ray radiation have studied the objects in the universe by the radiation they emit.

D. Advances in Biochemistry, Genetics and Molecular Biology--These areas are among the fastest growing of all scientific disciplines. The study of the biochemical properties of organisms through their DNA and genes is yielding benefits far beyond our greatest expectations. An outgrowth is biotechnology that uses genetic engineering to manipulate and modify natural organisms. Traits may be introduced into existing organisms to create new drugs, plants, diagnostics and medical and industrial processes. Biotechnology may turn out to be the biggest growth industry of the 21st century.

With genetically engineered bovine somatotropin, dairy farmers get more milk from their cows with less feed and waste. Goats are bred to produce human therapeutic and diagnostic proteins in their milk. Transgenic tomatoes that taste better, resist spoilage and grow cheaper, are now on sale. The genetically engineered yellow crookneck squash may be sold unlabeled at stores. Corn genetically engineered to produce pesticides against the European corn borer, has been approved for production by the Environmental Protection Agency.

Certain ethical questions need debate by an informed public. Should genetically altered animals and plants be patentable? Should a person with a genetic defect be allowed to remove that defect--and undergo treatment to assure that it is not passed on? Should research be permitted to improve the human species or to create human characteristics on demand? Should employers and insurance companies have access to an employee's genetic file?

THE NEED FOR INFORMED CITIZENS

The citizens of our nation must make scientific decisions every day that affect the public welfare. Some questions that have arisen recently in Southern California are typical of those confronting people all over the World. Is the public endangered by the transport of nuclear reactor fuel rods on freeway 101 and Amtrak through the city of Santa Barbara? Should the restrictions on the development of thousands of acres of kangaroo rat habitat in Riverside County be reduced or eliminated? Will it be possible or advisable to phase out and replace gasoline powered cars in Southern California with battery electric cars? When, if ever, will clean, cheap, plentiful alternative fuels and sources of electrical power be available?

Will the ban of many pesticides be of benefit to or will it harm the public? Should the endangered species act be modified? What should be the future of nuclear energy? Where will we store low-activity nuclear wastes from hospitals, universities and private labs? And where will the high-level wastes from spent reactor fuel rods be kept?

The proper decisions will come only from an informed electorate, from voters who can choose the best science and technology, realize the consequences of their choices and are willing to take action.

It is not at all clear that we citizens are prepared for the tasks ahead. Instead, many of us fail to register and do not vote. Only 55 percent of those of voting age cast ballots in the presidential election in 1992. And most who do vote are poorly informed in science.

In the United States in 1993, about 88 percent of the population 18 to 24 years old had completed high school and were no longer enrolled. Many graduated with no course in the basic sciences of biology, chemistry or physics. American high school seniors, in 1990, ranked last of 14 countries in an international science test.

One of the requirements for admission to a campus of the University of California is graduation from high school with grades in the upper 12.5 percent of the class. The students are required to have taken at least three years of math and two years of laboratory science. Yet it was clear in a campus-wide beginning astronomy course I sometimes taught, that many of the humanities students had little understanding of math and science.

A survey of 2000 American adults, recently released by the National Science Foundation, confirmed their lack of knowledge of science. In one multiple choice question adults were asked, "How long does it take for the Earth to go around the sun: one day, one month, or one year?" Less than half gave the correct answer, one year. If their answers were pure guesses, one-third would be correct by chance alone. When given the statement, "Electrons are smaller than atoms. (True or False)," only 44 percent gave the correct answer, True--even less than the 50 percent expected by chance. And when asked, "Tell me, in your own words, what is a molecule?", only nine percent knew the answer.

The results of a poll [1] of 1,236 adults showed that the belief of the American public in the paranormal is widespread. The paranormal are those events that cannot be explained by science. Half of the public believes in extrasensory perception; the ability to acquire knowledge of events, transmit knowledge to others and predict the future without using one of the five senses: sight, sound, smell, taste and feel.

One-quarter of Americans believes in astrology, that our lives are controlled by the positions of the moon and planets; and one-quarter in ghosts. About one-half of Americans believe that UFOs are real--that they are not misidentified stars, planets, airplanes, scientific balloons, lights or other normal objects. The hundreds of thousands of Americans claiming to have been kidnapped by aliens from space is even more disturbing. Not one piece of credible evidence has been found for either UFOs or kidnappings in the 50 years since they were first reported.

THE "SCIENCE VIEWS" COLUMNS

The "Science Views" columns have been gathered into six chapters of loosely connected groups of similar topics. They address problems that require knowledge of science for solution. Some are devoted to problems reported in the news by the media.

While it is not possible nor prudent to cover all issues, a large number of controversial subjects are discussed. Positions are taken on difficult questions, always with an attempt to give arguments supported by scientific evidence.

The columns are not infallible. Reasonable scientists can look at data and sometimes disagree. I think you'll find that these columns usually agree with conventional science.

CHAPTER I

SUPERSTITION AND THE PARANORMAL

P. T. Barnum once said, "There's a sucker born every minute." Although he was talking about the circus he could have just as well pointed to superstition and the paranormal. Superstition is a belief in magic, paranormal in the supernatural; neither can be justified by science. They are found even in the highest offices of the nation.

We're told that President Ronald Reagan's wife, Nancy, consulted an astrologer about safe times for the President's trips. From testimony released by the grand jury investigation of the Orange County bankruptcy, we find that County Treasurer, Robert L. Citron sought advice from psychics. Twenty million dollars of federal tax money was spent by the Department of Defense on futile attempts of psychics to locate Soviet submarines, North Korean hidden uranium, and stations of enemy spies.

We are flooded by tales of mythical monsters like the Abominable Snow Man, Big Foot, Loch Ness Monster and the recent Chupacabras, the Mexican Goat Sucker; cured of our ills by fraudulent faith healers with help from assistants mixing in with the crowd; and misled by diviners with forked sticks finding water underground no more often than is expected by keen surface observation and chance.

Claims of levitation by meditation, palm reading, special extrasensory perception, remote viewing, channeling and many others are believed and upheld by tens of millions of Americans. What happened to the skeptical public? Why aren't we questioning these beliefs that are unsupported by scientific evidence?

One strong group devoted to examining these claims is the Committee for the Scientific Investigation of Claims of the Paranormal (CSICOP). It is a non-profit organization that publishes the "Skeptical Inquirer", a journal enjoyed by many readers who want to expose the false arguments of the paranormal.

Among the charter members is James Randi, a magician and writer, who has exposed several scams including ones of the magician Uri Geller who claimed to bend spoons with his supernatural powers. Randi debunked a faith healer who fooled audiences by furnishing a wheel chair before the service to a visitor who was crippled, but could walk. During the service the healer dramatically cured the patient who then miraculously walked away from his chair.

Another is Carl Sagan, Professor of Astronomy at Cornell University, who is well known for his popular writing about science in his books, contributions to Parade Magazine and presentations in the TV series, NOVA. He has a few rules for critical thinking. (1) Propositions not testable are worthless. (2) There must be substantive debate, not attacks on the antagonists' characters. (3) Every link of the chain of argument must work. (4) Arguments from authority carry little weight.

Arguments must be supported by evidence. The more a claim deviates from accepted knowledge, the stronger the evidence must be. Always keep in mind, extraordinary claims require extraordinary evidence.

1. UFOs, Social Security and Generation X

The public's leading fantasy of the last half of the 20th century is the belief that Unidentified Flying Objects, UFO's, are spacecraft flown by aliens from outer space. Accompanying that is the disillusionment of the X generation with the government and its future.

The results of a September 1994 pole [1], "Surveying Generation X", reported that "while 46 percent of 18-to-34 year-olds believed in unidentified flying objects, only 28 percent believed that Social Security will exist when they retire."

The pollsters didn't distinguish whether those polled believed UFOs were expected events like unidentified blinking airplane lights and twinkling stars, or aliens from distant planets covered up by conspiracies of government officials. But the inference of the latter was clear. This comparison underlines the lack of understanding of science by the public and their loss of trust and confidence in elected government officials.

Gallup polls of adults over the last couple of decades have shown that the belief in UFOs by the public has stayed close to 50 percent.

Recently, "The UFO Coverup?--Area 51" was the subject of the 'Larry King Live' show on TV. It was yet another showcase for UFO activists who rehashed past events, some over 40 years old. The program contained the usual false claims of evidence for UFOs that could not be explained by natural or man made causes.

Their lack of ability to produce the required convincing evidence was blamed on cover-ups by government officials. Sensational best sellers of 'true' stories about kidnappings of people by aliens in the late 1980s, however incredulous, have not been subjected to the usual scrutiny by normally discriminating people.

During the 1970s and early 1980s, I, with my colleagues and students, flew more than 10 flights of scientific balloons at high altitudes for a day or two each over Texas and New Mexico. Almost always, reports of UFO sightings appeared in the local newspapers.

Some of the blame for the lack of criticism of pseudoscience certainly falls on the media, TV and press, who cater to an audience that prefers the sensationalism of UFOs to the exploration and reality of science. Exceptions are TV shows like NOVA that give the excitement of discovery in astronomy and biotechnology and medicine.

Most newspapers and magazines report stories straight, no matter how far-fetched, without comment or evaluation. This is a disservice to readers, many of whom have difficulty distinguishing between fact and fiction and sense and nonsense in science.

The Skeptical Inquirer is a magazine with critical articles about UFOs and other pseudoscientific subjects. Philip Klass, for example, has several articles about UFO sightings, where he points out the lack of evidence, inconsistencies and improper selection of or neglect of data.

The readers of newspapers would benefit immensely if experts like Philip Klass were consulted on columns concerning UFOs, or other scientists on columns in their areas of expertise.

There also seems to be a failure in teaching science to general students over the last 30 years in the grades through the universities. And scientists have not done a good job communicating the thrills of their discoveries to the public.

Many students have not learned to be skeptical--to test and question what they read and hear. The public does not have a sound understanding of science to cope with the large volume of nonsense that bombards it every day. The average citizen is not able to respond from his poor base of scientific knowledge.

The citizen should have the background to reply, "Travel to the earth from stars or galaxies is highly unlikely because of the great distances, long periods of time required and the great amounts of energy that would have to be expended. This is dictated by Einstein's Theory of Special Relativity. There is no way out. Therefore, before I believe an extraordinary story about UFOs, I will require extraordinary evidence. I will not be satisfied by a story about a landing of a bright light in a farmer's pasture with no evidence but a hole in the ground. Or by a piece of equipment picked up in the desert that looks like and is the remains of a weather balloon and its antenna. I will require hard evidence, the spacecraft, equipment from that craft or the aliens themselves. No excuses."

Wrong science is bad enough but UFology has also been a major player in encouraging distrust of government. A cover-up by government officials is blamed for their lack of definitive evidence for UFO landings.

This is ludicrous, as any person in or out of government would love to be the first to make this alien information public and any secret classification would be totally ignored. It would be the most exciting story of all time. It would be impossible to keep the information quiet.

Much of the loss of confidence in politicians has been brought on by themselves. Many are so intent on reelection that it controls their every action. Most of their campaigning is irrelevant.

Gridlock in the House and Senate has become intolerable. The cynicism of the 18-34 year olds is understandable--but not excusable. Their voting record is the poorest of any age group. They have had better education, health and opportunity than their predecessors. They will determine the government of tomorrow and need to make their political clout felt today.

With the leadership of the teachers and professors of our schools and universities, the cooperation of responsible TV and press, answerable government officials, responsible young adults, and the necessary support of citizens of all ages, future polls could find the percentages reversed.

The time could come when only a few percent of Generation X believe the fiction of UFOs and nearly 100 percent believe their future retirement is secure.

2. Alien Abductions

Promoted by sensational alien abduction books, nurtured by 'abduction recovered memory' experts and fueled by superficial talk shows, "alien abduction" has been the mass psychosis of the last two decades.

It is estimated that more than a million people believe they have been abducted by aliens and millions more think they have been contacted.

This without a scrap of credible evidence. No downed saucers, spacecraft parts or alien corpses. Only blurred photos, indistinct videos and several hundred thousand anecdotes.

This whimsical fancy was reviewed again recently in "Kidnapped by UFO's" which aired on many Public Broadcasting Stations. It started with the tale by Betty and Barney Hill of their encounter with aliens on a lonely road in the 1950's.

In later encounters people told stories of aliens with thin gray bodies, big black eyes and large egg shaped heads. They walked through walls, lifted people upward by light beams to flying saucers. They operated on earthlings, probed their ears and anuses and implanted devices up their noses.

The strange creatures reportedly induced erections in men and removed their semen. One woman related that an alien fetus emerged from her body. These stories were accompanied by appropriate "alien background" music.

Considerable TV time was given to Bud Hopkins, writer of abduction thrillers, an "abduction recovered memory" expert and organizer of alien abduction support groups. After watching the film of one of Hopkin's therapy sessions with a four year-old boy, psychologist Elizabeth Loftus observed that Hopkins was implanting memories rather than recovering them.

Even Harvard University is not immune to the alien abduction epidemic. John Mack, professor of psychiatry at the medical school, wrote the best seller called, "Abduction: Human Encounters with Aliens." It described 13 case studies of patients who claimed they were kidnapped by space aliens and carried away in flying saucers for sexual experiments.

Mack appeared on "The Oprah Winfrey Show", "Larry King Live" and other talk shows. But it seems Harvard still has its cross to bear. After a year's deliberations, the typical decisive academic review committee found that Mack had violated no university rules in his research with alien abductees. And since he has tenure was free "to study what he wishes and to state his conclusions without impediment."

Nevada recently designated a 92 mile-length of Highway 375 as the 'Extraterrestrial Highway'. At the "Little A'Le Inn" in Rachel, Nev., terrestrial tourists can dine on 'alien burgers' and buy UFO T-shirts and bumper stickers. UFO sightings are not guaranteed.

Nor is Santa Barbara to be denied. It has the Institute for the Study of Contact with Non-Human Intelligence. This includes aliens and angels. UFO's are a core curriculum. Michael Lindemann, President and co-founder of the Institute, in the 1980s was previously executive director of Santa Barbara's Peace Resource Center.

The public infatuation with visits from outer space can be traced back to Erich von Daniken's "Chariots of the Gods?". It was published in 1969 and sold over 4 million copies. His spacemen found a primitive species of humans on earth. Artificial insemination led to our Homo sapiens species of today.

Where is the skepticism of the American public? And the demand for tests and proof? Many adults seem to have fewer doubts about the little green men than most three year-olds about Santa Claus or the Tooth Fairy.

Alien abductions have absolutely no support from scientific observations, tests or theories. Even more damaging, the reports of the activities of millions of these creatures and their spacecraft from outside the solar system violate well-established laws of physics.

Extraordinary claims require extraordinary evidence. After 40 years of preposterous unsupported claims it is now time to admit that alien abductions are nothing more than vivid imaginations, dreams and hallucinations of people easily influenced by suggestion.

The public does not need this space-age religion with its demigods from outer space.

Instead psychologists, psychiatrists, social scientists and physicians need to study the causes of this mass hysteria and devise appropriate treatments for the abductees and their problems.

3. Astrology

Three thousand years ago the myth of astrology seemed reasonable to the ancients. Today with our vast array of large astronomical telescopes and sophisticated knowledge of the sun, planets and the moon, astrology makes no sense. It is no more supported by science than are the gods of the early Greeks.

As President Bill Clinton and Senator Bob Dole faced off in 1996 preelection debates, we were reminded that Nancy Reagan selected 'good' days for then President Ronald Reagan's travel by consulting a San Francisco astrologer.

Although this was never a campaign issue for President Reagan, columnists and humorists had a field day. Compared to many trivial non-issues that saturated TV and the newspapers during the elections in the 1980s, astrologers as advisors to politicians could have rated national importance.

It is of no surprise that a 1991 Gallup Poll found that one-fourth of American adults believe our lives can be controlled by positions of the stars and planets, 75 percent read their horoscopes occasionally and 25 percent at least once a week. This is little change from a similar poll taken in 1979. Superstitions and astrology are probably more discussed in personal conversations than all other paranormal subjects.

History tells us that astrology originated among the Mesopotamians and Babylonians about 1,000 years BC. They believed that the motions of the planets influenced the fortunes of kings and nations.

The Greeks, by the second century BC, had added the notion that the planets affect the lives of all people, independent of status. Planets were named for the gods and were given powers the gods possessed.

This natal astrology reached its apex in the second century AD with the publication of Ptolemy's "Tetrabiblos," the 'bible' for astrology since then.

The horoscope charts the positions of the planets at the time of an individual's birth. The known planets, plus the sun and the moon, also thought to be planets, were located on the zodiac that was divided into 12 sectors called signs.

The orientation of the celestial sphere, on which the stars were considered fixed, was also specified at the time of birth. The celestial sphere was divided along the horizon into 12 houses, and each day the planets and signs moved through the houses.

Interpretations of horoscopes by astrologers furnished advice to clients and guided their future. Horoscopes appear in most newspapers and many carry warnings that the astrological forecast should be read for entertainment only, its predictions have no reliable basis in scientific fact.

The ancients didn't have the advantages of our day. We have 400 years of observations with optical telescopes, 50 years with radio telescopes and some 30 years with infrared, ultraviolet, x-ray and gamma ray telescopes observing above the atmosphere from high flying scientific balloons, rockets and satellites.

The elders didn't have the advantages of analyses of samples of material from the moon and remote sensing by instruments on satellites in orbit around and passing close to the planets. They didn't realize that the planets are made of material similar to the earth.

Newton's Law of Gravitation that holds the earth and other planets in orbit about the sun, and moons in orbits around the planets, is now well understood. We know that the gravitational forces of the planets, other than the earth, on the child at time of birth are minuscule compared to that of the attending doctor, or the hospital bed.

The forces of nature are known. No force from the planets could affect human lives. Reflected sunlight or other radiations from the planets pale beside light from the sun and could have no consequence. Other radiations from the sun are absorbed by the earth's atmosphere. And of course, the walls of the hospital absorb even the sunlight that reaches the surface of the earth.

Today, we know there is no basis for supernatural beliefs of astrology. Why then the continued belief of 25 percent of the public?

Much of the interest is recreational. "Oh, dad, it's just for fun." Most followers are casual. Horoscopes give simple solutions to complicated problems. The advice comes in bite sizes.

I'm a Capricorn. On November 3, 1994, one horoscope counseled, "Take advantage of continuing trends that favor socializing and romance. Avoid a dispute over money." Another on the same day cautions, "Family members pull you in two directions--remain with older individual whose loyalty is beyond question."

Couldn't anyone, Aries to Pisces, see a part of themselves in these quotes? Christopher French and colleagues suggest that the Barnum effect provides the best explanation for belief in astrology[2].

P. T. Barnum, the famous circus owner, attributed his success to two considerations--always have a little something for everyone, and there's a sucker born every minute.

Even with the best of intentions, astrological advice to some people could lead to disastrous consequences. And, when applied to the President's schedule of meetings, could negatively affect world events.

Rather than playing with horoscopes, why not learn about the new discoveries in astronomy and with friends discuss the latest ideas about the origin and evolution of the universe? Such activities could be more enjoyable and satisfying than continuing the superstitions of astrology practiced long ago by the ancients.

4. Astral Projection

Astral Projection is the ultimate fantasy--more fun than coasting down steep hills on a bike, or sailing through air on a glider. Why not out-of-body travel to an exotic vacation resort? Too bad, it's only a dream.

I was aghast on turning to the Life section of our local newspaper. Under a SCIENCE header was a feature story [3] about 'astral projection'.

For the uninformed, astral projection is out-of-body travel and adventure. Nancy Trivellato, of the International Institute of Projectology in Miami, we are told, came to Florida to set up the new Institute and is holding seminars for those interested in science, not mystical ideas.

Trivellato was a former administrative assistant at IBM in Brazil who attended a class at the Institute 4 years ago and then became an instructor. The Projectology instructors say you can learn the tricks to 'soul tripping' in a few easy lessons.

Why the science section for this feature? What is the science involved? No double-blind tests were suggested or experiments proposed. Only anecdotes were offered.

Is this not better suited for the entertainment section where amusing stories are expected? Or recreation since travel by astral projection is easy without need for auto, train, airplane or ship? Because of its mystical qualities and since belief is by faith, could it appear on the religious page?

The astral projections remind us of arguments of reincarnation given in "Twenty Cases Suggestive of Reincarnation" by Ian Stevenson in 1974. This book has been influential for the believers but is criticized by Leonard Angel and others [4].

Similar are the 'out-of-body' experiences into death claimed by patients while lying apparently unconscious on their hospital beds. These would be ordinary dreams except for information they report, that some say, shouldn't have been available to them.

The near death experiences are in many ways reminders of the outrageous alien abduction stories that resurfaced in the 1990s. These have been fueled by a privately funded Roper Survey using 11 questions constructed by Budd Hopkins and David Jacobs, promoters of stories that aliens from outer space have abducted earthlings. The claims are questioned by Lloyd Stires and by Philip Klass [5].

The Roper survey didn't ask the 5,947 adults if they were UFO abductees; only less threatening questions like "How often has this occurrence happened to you?", followed by "Waking up paralyzed with a sense of a strange person or presence or something else in the room."

Five of the questions were considered 'key indicator' questions by Hopkins and Jacobs, the other 6 only neutral. Just 18 or 0.3 percent of the participants marked all 5 of the key indicators. Extrapolating to the 185 million adult population gives 560,000 adults abducted.

Hopkins and Jacobs then relaxed their criteria requiring only 4 out of 5 of the key indicator questions to be positive. This gave the much publicized 3.7 million Americans abducted by aliens. Assuming that the kidnappings began with the abduction of Betty and Barney Hill 30 years ago, one reaches the absurd conclusion that an adult is abducted by aliens every 4 minutes.

There are numerous flaws in the questionnaire. One is that no control population of American adults, who claimed to have been or not have been abducted by aliens, was asked these same 11 questions.

But the biggest flaws are claims of kidnappings by extra terrestrial aliens of Americans all over the United States without photographs, videotapes or even scraps of evidence. Where are the levitation devices and metallic spacecraft in which the victims were transported and subjected to medical examinations? And where are the aliens and the hybrid race produced by impregnating women?

Why don't these intelligent aliens contact the heads of state, or NASA or the scientists in the research universities?

The strongest evidence of all against this nonsense, orchestrated by "Oprah", "Larry King Live", "Hard Copy" and "Unsolved Mysteries" is the violation of physical laws including the Special Theory of Relativity required to bring the aliens from planets around distant stars to the earth and return.

5. Psychics

We wonder why people ever consult psychics about future events. If psychics really have extrasensory perception they wouldn't bother with my question about whether tomorrow is a good time to fertilize the roses. They would all be at the race track becoming millionaires

These are heady days for the psychics. In November, 1995 the Defense Intelligence Agency revealed that psychics, claiming paranormal power, had been employed frequently over the last 20 years in their 'Stargate' program.

Twenty million dollars of our federal tax money was spent consulting psychics. They were asked where North Korea was hiding its uranium and where it had dug tunnels under the demilitarized zone toward South Korea. Psychics were requested to locate the stations of enemy spies and determine the assignments of members of criminal organizations.

The negative results should have been predicted by the Department of Defense--as well as by the psychics. A review panel recommended dumping the program.

In testimony recently released from the Grand Jury investigation of the Orange County bankruptcy, we learned that the former County Treasurer, Robert L. Citron, sought advice from psychics. Huge county losses of $1.64 billion were announced; Citron resigned.

Ted Turner launched his new TV business channel, CNNfn, with "The Amazing Kreskin", a psychic who predicted a slow but steady growth for the channel.

As a demonstration of chance, in selecting stock market winners, portfolios chosen by pins stuck randomly into the stock page sometimes surpass the average chosen by experts. So the psychics are expected, by chance, to do as well, too.

Robert Sheaffer, in his "Psychic Vibrations" column [6], gives an example of Kreskin's skills. Amazing Kreskin wrote to Mrs. Dorothy Lea of British Columbia, "I come to you, Dorothy Lea....I am going to use my power to put you in a world.....of big money." But first she had to send in $25 Canadian.

At the bottom of the letter Kreskin asked her to fill in her date and year of birth--a request by this psychic who claimed to read minds and tell tomorrow's headlines. The letter was a good test of Amazing Kreskin's psychic skills. Mrs. Lea had been dead for 2 years.

Psychic advice is as close as your telephone. For $3.99 per minute you can talk to a psychic on the Psychic Readers Network. They get 25,000 calls per day. The Network claims this is only entertainment. But it doesn't cost $240 an hour to rent a video or see the latest movie.

William Grey, Professor of Philosophy, University of New England, Australia gives an excellent review of the philosophy of the paranormal [7]. In it he discusses Psychokinesis (PK), Extrasensory Perception (ESP) and a third area of the paranormal consisting of out-of body experience, disembodied existence, reincarnation and life-after-death [6].

PK is the ability to affect physical events by thought without the intermediary of bodily action. ESP can be divided into telepathy, clairvoyance and precognition. Clairvoyance is the ability to acquire knowledge of events through other than the known senses. Telepathy involves transfer of knowledge from person to person and precognition, knowledge about the future.

Physicists have discovered four fundamental forces that govern all interactions in the universe:

The gravitational force--my weight is caused by the gravitational pull of the earth; and the motion of the earth in its orbit by the gravitational pull of the sun.

The electromagnetic force--chemical reactions, magnets, radio, TV and computers depend on electric and magnetic interactions.

The strong or nuclear force--strong interactions among neutrons and protons are required to hold nuclei of atoms together and is important in fission and fusion.

The weak force--is responsible for certain decays of nuclei and particles but plays no role in the interaction of normal matter.

No other forces or interactions have been discovered in nature. So any beyond the four would be considered paranormal. This does not mean that other forces or interactions are impossible. Only that none have survived rigorous scientific tests. Thus any claim of paranormal behavior by psychics of clairvoyance, telepathy or precognition must be supported by extraordinary evidence.

To date, none have survived the test.

6. Magicians, Randi and Geller

The two magicians, James Randi and Uri Geller, could not differ more. Geller performed magic tricks which he claimed were paranormal. Randi duplicated the tricks and showed they were the normal slight of hand of a clever magician.

Uri Geller, a magician-psychic famous for bending spoons and claiming paranormal powers has lost another round to James Randi, the magician-skeptic.

For the last 25 years Randi has exposed Geller's bending feats and other tricks as those of a magician, only, and his extrasensory perception as the fraudulent deception of a confidence man.

Geller bent spoons on the table while distracting the audience with his chatter. His knowledge about strangers in the audience came from observing them in the parking lot or by signals from a colleague planted in the front row.

Randi was quoted as saying that even reputable scientists had been conned by Geller's tricks that "are the kind that used to be on the back of cereal boxes when I was a kid. Apparently, scientists don't eat cornflakes anymore" [8].

In May 1991, Geller filed a $15 million suit in the U.S. District Court against Randi. The Committee for the Scientific Investigation of Claims of the Paranormal, CISCOP, was also named.

Eventually CISCOP requested the court to end the lawsuit and impose sanctions on Geller for bringing it. Barry Karr, CISCOP Executive Director said that CSICOP was made a defendant solely to harass and intimidate with the hope that the lawsuit would discourage discussion and analysis of paranormal claims, especially those of Geller.

In July 1993, U.S. District Judge Stanley Harris ruled that Geller should pay $149,000 to CISCOP for its court costs. After a challenge by Geller, in December 1994, a three-judge appeals court upheld the earlier ruling.

Geller was born in 1947 in Israel. According to his autobiography, his spoon bending began as a boy, while eating mushroom soup. The spoon he was using suddenly bent, spilling the soup onto his lap.

Golda Meir, Israel's Prime Minister, made him famous in the late 1960s when she replied to a radio questioner about Israel's future, "Don't ask me--ask Uri Geller."

Russel Targ and Harold Puthoff of the Stanford Research Institute (SRI), Menlo Park CA, invited Geller for a series of tests on remote viewing. The best known are the 13 extrasensory perception experiments. Geller drew pictures in the "subject enclosure", of drawings chosen at random in "the experimental area." Targ and Puthoff reported Geller identified seven of 13 targets. A result far above that expected by chance.

These results [9] were published in the scientific journal "Nature." Although the paper had been rejected previously by two other journals it was accepted with several reservations. Among them: the referees agreed that the paper was weak in design and presentation; and that safeguards and precautions against conscious or unconscious fraud was "uncomfortably vague."

Nature said the paper "does not present any evidence whatsoever for Geller's alleged abilities to bend metal rods by stroking them, influence magnets at a distance or make watches stop." Nature argued that publication allowed scientists the opportunity to repeat and evaluate the experiments. It didn't give its seal of approval to the results.

Randi was born in Toronto, Canada where his magic was self taught. At age 15, he saw a preacher reading the contents of sealed envelopes and called the spiritualist a fake. Following the resulting fracas he spent 4 hours at the police station. At 17 he dropped out of high school and joined a carnival as a magician.

He is a well known skeptic, helped organize CSICOP in 1976 and 10 years later received a prestigious 5-year MacArthur Foundation Fellowship. For 20 years he has offered $10,000 to anyone who can demonstrate paranormal ability under controlled conditions. Of 600 applicants and 75 serious tests, no one has succeeded.

Randi's books include "The Magic of Uri Geller" published in 1975, "Flim-Flam!" in 1982 and "The Mask of Nostradamus" in 1990. In the first two, Randi points out many flaws in the Targ and Puthoff experiments. They were sometimes guilty of selective reporting, keeping tests that were successful and discarding those that failed. The subject room where Geller made his drawings had a 3 1/2 inch hole stuffed with gauze that connected to the experimental area. Conversations between Targ and Puthoff could be heard by Geller.

Contrary to the "Nature" paper, many of Geller's drawings were not reported.

The tests at SRI never showed that Geller could perform paranormal feats as he claimed. And when given the opportunity to deceive, he usually did.

7. THE CHUPACABRAS

Every culture has its grotesque monsters. In Greek legend, the Cyclops were one-eyed cannibal giants that lived in Sicily. The Chinese celebrate with dragons. Mexico's latest frightening monsters are the Chupacabras.

For weeks the Mexican media have been filled with terrifying tales of the blood-thirsty "chupacabras," translated, the goat suckers.

A man in Sinaloa, a Pacific coast state, claimed seeing a fierce bat-winged beast swoop down on his flock. He found 24 sheep dead with vampire-like bites and their blood sucked dry. Immediately, deaths of chickens, goats and sheep all over Mexico were blamed on the goat sucker.

Farmers waited for daylight to go to the fields and children stayed home from school. At the Governor's ranch in Guanajuato, an employee said he saw the beast, a bat-like creature with fangs, bug eyes, wings and kangaroo legs.

Two goat-sucker world-wide-web pages were opened on Internet. Intellectuals critiqued chupacabras over cocktails and lunch. And peasants in panic formed posses to hunt down the beasts.

Scientific reports and autopsies blamed starving packs of coyotes and wild dogs. In Sinaloa, 15 local university biology faculty and state zoo experts along with 25 members of the police swat team organized traps. Wild dogs attacking a corral of sheep were captured and shown to the citizens. Other traps were successful at several localities. Still the people were not convinced and the rumors keep flying.

Vampires and humanoid monsters have been legend since the beginning of recorded history. In spite of our vast knowledge from studies in anthropology, biology, zoology and the many other sciences, an abundance of myths, folklore, and fantasies still abound.

The "Yeti", or "abominable snowman", according to sherpa legend, inhabits the lower slopes of the Himalayas. He's a mythical monster that leaves tracks in the snow.

Many of the sherpas live high in the Himalayas. Thousands of trekkers climb the trails every year and several expeditions have hunted for the Yeti. Yet no one has seen one, dead or alive.

The marks in the snow, not caused by falling stones, are probably produced by bears. At certain speeds bears' hindfoot prints overlap forefoot prints, and make very large tracks that appear to be enormous human footprints traveling in the opposite direction.

The first report in Scotland of the "Loch Ness Monster", called "Nessie", appears to be that of Saint Columba who stopped Nessie from devouring an unfortunate onlooker by evoking the name of the Lord. Scientific evidence apparently goes back to 1960. The earlier much publicized "surgeon's photograph," taken in 1934, was later found to be fake.

Rikki Razdan and Alan Kielar, electrical engineers and specialists in sonar and image processing at ISCAN, Inc. in Cambridge, Massachusetts, searched for Nessie with "sonar," sound waves, and evaluated previous sonar experiments [10].

Their monitoring for seven weeks to a depth of 100 feet in Loch Ness found no evidence for a fish longer than three feet. Their analyses of previous measurements revealed discrepancies that could not withstand the new scrutiny. No valid evidence now remains for Nessie, the Loch Ness Monster.

In the Pacific Northwest, we have our own folklore of "Sasquatch", or "Bigfoot," a humanoid monster who leaves giant footprints in the ground. Michael Dennett, a veteran investigator-writer evaluated the evidence for Bigfoot [11]. It consisted of footprints from the Mill Creek watershed in the Blue Mountains of Oregon found in June 1982, the "Cripple Foot" footprints found in 1969 near Bossburg, Washington, two sets of handprints and one footprint from Bloomington, Indiana.

Dennett discovered that the "Bloomington track" was an admitted fake. Most of the other tracks were found by Paul Freeman, at that time a new Forest Service patrolman. He also claimed finding samples of Bigfoot's hair, making a tape-recording of its screech, its photograph, and twice encountering it face to face. The hair was analyzed and found to be artificial.

Representatives from the Forest Service were convinced that the Mill Creek tracks had been hoaxed.

Unless more compelling evidence is found, the Chupacabras will remain a myth joining the fables of the Abominable Snowman, Loch Ness Monster and Bigfoot.

8. The Shroud of Turin

New techniques in age dating have made it possible to determine the ages of many art, historical and religious objects with precision. Occasionally one, such as the Shroud of Turin, is found to be counterfeit.

The Shroud of Turin is reputed by many to be the burial cloth of Jesus Christ. Photographic plates exposed in 1898, suggested negative images front and back of a human body with wounds. This was presented as evidence, in 1902, to the French Academie des Sciences as the imprint, not a painting, of the body of Christ.

The Shroud can be traced back only as far as the 1350s where it was on exhibit at Lirey, France. It went to the Dukes of Savoy, and after several stops, to Turin in 1578. It was placed in a special shrine in the Royal Chapel of the Turin Cathedral in 1694.

Scientific examination was first permitted in 1969 and 1973 by a committee appointed by Cardinal Michele Pellegrino and then again by The Shroud of Turin Research Project in 1978. High priority was given to the determination of the age of the flax in the linen cloth.

The method of dating material from plants and animals, still today, is carbon-14 decay, developed by Williard Libby, who received the Nobel Prize in chemistry in 1960 for his pioneer work. The addition of accelerator-mass spectrometry and small gas detectors in the 1970s made possible highly sensitive analyses with small pieces of sample.

In October 1987, labs at the Universities of Arizona, Oxford and Zurich were selected by the Archbishop of Turin, acting on instructions from the Holy See, to carry out the measurements. Procedures and results of the

Radiocarbon dating of the Shroud were published by P. E. Damon and 20 authors [12].

The earth's atmosphere is about 80 percent nitrogen and 20 percent oxygen. Collisions of cosmic ray protons with atmospheric nuclei produce neutrons. When a neutron is captured by a nitrogen-14 nucleus, a proton and a carbon-14 nucleus are formed.

The carbon-14 is radioactive with a half life of about 5,570 years. It decays back to nitrogen-14 by emitting an electron. The half life is the time for one-half of the nuclei to decay.

Growing plants absorb carbon dioxide during photosynthesis. Most of the carbon is the naturally occurring carbon-12 and carbon-13 in abundances of 99 and 1 percent; a very small fraction is the radioactive carbon-14.

This mixture of the 3 isotopes of carbon in the atmosphere is nearly constant over tens of thousands of years. When the plants die, photosynthesis stops and they no longer take in new carbon dioxide. As the carbon-14 decays, its abundance relative to carbon-12 and -13 decreases. The relative abundance of carbon-14 then determines the age of the sample.

The decay rate of the sample can be measured with a low energy electron detector. As the sample ages the decay rate decreases. From the number of carbon-12 nuclei in the sample and the efficiency of the counters for detecting the decays, the age can be determined. This method was used by Libby and his followers.

A more sensitive way to determine the age is to use accelerator mass spectroscopy. This method was necessary for the Shroud because of the very small samples available to the labs. At the labs the samples were combusted to gas, then deposited on targets. The sample atoms were ionized and focused into the accelerator where their energies were increased to a few million electron volts.

They were then passed through magnets that separate the 3 isotopes and are counted on 3 different detectors. Here, a large fraction of the nuclei leaving the target are counted instead of just the ones that decay. The ratio of the numbers of carbon-14 nuclei to carbon-12 and -13 from the sample compared, with the ratio in the atmosphere, after appropriate corrections for efficiencies, gives the age of the sample.

Three samples of known ages were also run as controls. One linen was from a tomb at Qasr Ibrim in Nubia dated from the 11th to the 12th centuries

AD, a second from a 2nd century AD mummy of Cleopatra from Thebes and a third from a cope of St. Louis d'Anjou dated about 1290 to 1310 AD.

The results from each lab were sent to the British Museum Research Lab for statistical analysis.

The results of all the work were combined to give a calendar age for the linen from the Shroud of Turin as 1260-1390 AD with 95 percent confidence. These results provide compelling evidence that the linen of the Shroud of Turin is medieval.

9. Noah's Ark

Adventurers need not have searched for remains of Noah's Ark on Mt. Ararat. There is not enough water on earth to float the ark that high. If all the ice on the earth melted it would raise sea level less than 300 feet.

The search for Noah's Ark on Mt. Ararat in Turkey has challenged adventurers from medieval to modern times.

Mt. Ararat--16,985 feet high--is capped by a glacier 200 square miles in area and 200 feet deep at heights above 13,000 feet. According to legend, a 12th century monk brought down pieces of wood found on the glacier, that were claimed to be part of the ark. Those artifacts were apparently destroyed about 1840 with the monastery where they were kept.

More recently, in 1969, a French industrialist and explorer, Fernand Navarra, found wood in a crevasse in the glacier. He sent pieces to labs around the world to determine their age. If part of Noah's ark, they were expected to be about 5000 years old.

The results from carbon 14 age dating, carried out independently at UC Los Angeles by Ranier Berger and at UC Riverside by Ervin Taylor, found ages of 1210 and 1230 years, clearly not old enough to be from Noah's Ark. Instead the wood may have come from a shrine erected by early Christians.

But the inescapable reason that a search at high altitudes was fruitless, is the lack of sufficient water to cover the land surface of the earth to such heights. The additional water available then, as now, is frozen as ice in the Antarctic and Greenland with lesser amounts in the arctic, glaciers and mountains. If all the ice on the earth were to melt, the sea level would rise only about 260 feet.

Most of the rugged area in extreme eastern Turkey surrounding Mt. Ararat is thousands of feet above sea level. If the ark were built at an altitude higher than about 300 feet above sea level, water would never have risen that high and the ark would never have floated.

The last ice age ended about 10,000 years ago. The climate has been gradually warming since then. Glaciers have receded. Average temperatures are warmer today than 5,000 years ago. Minor climate fluctuations have occurred. A warmer period happened about 1000 AD and a little ice age during the late 1600s.

From analyses of cores removed from Greenland ice sheets and from drilling into lake and ocean bottoms, climate changes over times of 10,000 to 100,000 years are mostly understood. The Milankovich theory, proposed in 1926 but not accepted by most climatologists until 1970, explains the variations.

Milankovich proposed that the climate depends on several properties of the earth's axis of rotation (which causes night and day) and the orbit of rotation of the earth around the sun.

The earth's axis of rotation does not keep the same direction in space but precesses like a spinning top. It takes 26,000 years for the axis to make a full revolution.

The major axis of rotation of the earth in its orbit about the sun spins in the opposite direction. About every 20,000 years, like now, the earth is closest to the sun during winter in the northern hemisphere. Then the northern winters are relatively mild and southern ones more severe. In ten thousand years the conditions will be reversed.

The earth's orbit around the sun is an ellipse. It is nearly a circle but its ellipticity (deviation from a circle) varies over time. This period is about 100,000 years. Also the angle between the planes of the earth's orbit around the sun and the earth's equator vary from about 22 to 24.5 degrees. The present angle is about 23.5 degrees. This variation has a period of about 40,000 years.

Sunspots are dark regions on the sun. They vary over time with average periods of 11 and 22 years. Many attempts have been made to associate weather time changes with the variation in the number of sunspots. Success has been limited.

An asteroid collision with the earth 65 million years ago placed large quantities of dust in the earth's atmosphere for tens of years that caused the deaths of the dinosaurs. Large volcano eruptions put lesser amounts of dust into the atmosphere but may cause weather changes detectable for years.

C. D. Keeling, an oceanographer at Scripps Institution of Oceanography at UC San Diego, and co-workers observed that carbon dioxide in the earth's atmosphere at Mauna Loa, Hawaii, increased continuously over the last 38 years. But has appeared to level off in recent years. It is caused by burning fossil fuels and biomass. Other greenhouse gases such as nitrous oxide and carbon monoxide are higher than preindustrial concentrations. Chlorofluorocarbons have increased solely due to human activities.

Global warming of 1 to 4 degrees Fahrenheit is expected by 2050. Precipitation will increase and sea levels are expected to rise 2 to 15 inches by 2050 compared to 2 to 5 inches over the past century.

CHAPTER II

SCIENCE AND ANTISCIENCE

The colleges and universities in the United States graduate the best trained scientists in the world. Students around the globe are attracted to our excellent graduate schools. Our scientific research and technology are the envy of the most advanced nations. Scientists in the United States have won more Nobel prizes in chemistry, physics and physiology-medicine than in any other country. Yet we don't do a good job of educating the public in science.

Less than half of U.S. students have more than one year of science in high school. In a recent international science competition of high school seniors, the Americans ranked last of 14 countries entered. On a National Science Foundation science survey in 1996, only 27 percent of the adults answered seven out of ten questions correctly, enough for a passing score.

Over 60 percent of industry representatives surveyed by Bayer Corporation said that most young adults are unprepared for industry entry-level jobs. They would prefer much more science training in the schools. More than 80 percent thought the ability to understand newspaper articles about science, for example, should be a requirement for employment.

Not only is the lack of science training a problem but some fraction of the population is opposed to science and technology. The Luddites, who smashed and burned factories in England nearly 200 years ago, symbolize the opposition. This spring, 1996, present day Luddites traveled by car and air--not horse and buggy--to their convention in Barnesville, Ohio. While non-violent these days, the Luddites still protest technology in every day life and the need for computer literate children.

In the last few years the attack on science has come from a surprising quarter--from what life Scientist Paul Gross and mathematician Norman Levitt in their book, "Higher Superstition" [1], call the 'academic left'. It embraces postmodernism, feminism, radical-environmentalism, multiculturalism and most recently social construction. These movements seem to share a dislike for scientists and scientific knowledge and methods. They do not agree that science is a body of knowledge that can be tested by experiments and observation.

Robert L. Park, Professor of Physics at the University of Maryland and Director of Public Affairs for the American Physical Society, has been an outspoken critic of the anti-science movement. In a contribution to the New York Times in July 1995 Park pointed out that environmental activist, Jeremy Rifkin, predicts disaster and organizes opposition to each new technological success. Rifkin's latest target is genetic testing. He launched his new attack in May (1996) against the patenting of breast cancer genes and lobbied for laws to limit access to human genetic data in research, medicine and commerce.

Sandra Harding, a professor and feminist at the University of Delaware claims the laws of physics were constructed to maintain white male dominance, and compares traditional methods of science to "marital rape, the husband as scientist forcing nature to his wishes."

Bernard Ortiz de Montellano, an anthropologist at Wayne State University, describes how destructive misinformation in science can be [2]. In 1987 the Portland, Oregon school system published a series of "African-American Baseline Essays," including one written by Hunter Adams, an environmental technician at Argonne National Labs.

According to Montellano, the essay was a jumble of myth, religion and folk medicine and "is a classical example of pseudoscience, but because of the current pressure on school districts to incorporate multicultural material in the classroom...it has been widely distributed." The baseline essays "have been adopted or seriously considered" by a number of school districts, including Detroit, Ft. Lauderdale, Atlanta, Chicago, and Washington, D.C.

The new national standards for teaching American history, prepared by dozens of historians working for three years, was released in 1994. Park notes that the new curriculum for grades 5-12, in the 250 page document, made only one mention of science. It appeared in a list of professions from which women had been systematically excluded. Equally upset was the U.S. Senate that condemned the report by a 99 to 1 vote. The National Center for History in the Schools, at UC Los Angeles, has rewritten the report. It has added a section on the major discoveries in science and technology and how they have affected the economy and daily life. The latest set of standards now seems agreeable to all.

The National Broadcasting Company (NBC) aired a TV program, "The Mysterious Origins of Man" in February, 1996, hosted by Charlton Heston. It claimed that evolution is a questionable theory, that human civilizations began more than 100 million years ago and that a conspiracy of scientists is suppressing significant archaeological evidence. Scientists [3] protested that the program contained pseudoscience that had misled the public. Instead of apologizing and withdrawing the program, NBC blatantly scheduled the show again on June 8. Bill Cote, the show's independent producer, explained, "NBC's extensive legal department put us through the wringer until we presented a balanced view."

Our youth and the public need to know that the universe is controlled by natural laws. The physical laws are universal, the same for all people in all places--independent of race, religion or gender. Scientists and the public have tested these laws millions of times under all sorts of conditions. Hundreds of times a day, each of us tests Newton's laws of force and of gravitation--every time we walk, run, and jump, or accelerate and stop our cars.

Only when Newton's laws were tested under very different conditions, at extremely high velocities near the speed of light and in the vicinity of very large masses like the sun, were they found inadequate. The laws were improved to satisfy these new conditions by Einstein with his Special and General Theories of Relativity. However, the changes in Einstein's theories have no effect on our daily lives. At the velocities and masses of our daily interactions with nature, Einstein's theories reduce to the familiar Newton's laws.

1. Introduction to "Science Views"

An editor at the Santa Barbara News Press twisted my arm, "We need a column to introduce you and 'Science Views' to the readers." So here it is.

In the following weeks, I will be writing a column for the News Press that will discuss science issues facing the public. But first let me introduce myself.

University professors of physics, even emeriti, usually write about their recent experiments or latest theories but seldom about themselves; except, of course, when preparing their files for merit increases and promotions.

My first 21 years were spent in Kansas, in small towns. My father was a school teacher, principal and superintendent and my mother a teacher before starting a family and after we children were grown. From an early age I owe much of the excitement of learning to my mother and father.

When asked about my home town, I always claim Pretty Prairie because that usually enhances the image most have of Kansas. I graduated from high school there. It was during the depression of the 1930s and my parents and two brothers and sister and I lived on the farm that year.

Three of us drove a model T Ford around the section to pick up a full load--then 7 miles to school. Among my courses were two semesters of physics. The teacher managed to get one quarter of the way through the book by June. But I loved physics so I read the rest of the text on my own.

It opened up an entirely new world. It explained why balls drop after being tossed up, why the sunset is red and how the rainbow is formed.

It explained electricity, how it's generated, and motors, batteries and radios. We didn't have TV, VCRs, computers or compact discs in those days. Jet airplanes were still on the drawing boards and nuclear physics was in its infancy.

One student from each school could enter a state-wide academic test in high school subjects. I chose physics and scored 97th out of 117 taking the test. It was a jar to my ego but not my ambition.

With a scholarship I went to a small Southwestern College in Winfield, Kansas, where I majored in physics, chemistry and math. I was fortunate to have two professors who took a personal interest in my progress. One taught physics and was Dean of Men.

It was to my advantage to be friendly with his student secretary. After graduation we were married and have 3 children and 7 grandchildren.

We went to the University of Illinois for my M.S. degree and on to the Navy as an Ensign where I was assigned to the Naval Research Lab to work on separation of uranim-235 from uranium-238 by thermal diffusion, Navy's only program in the Manhattan Project to develop the atomic bomb.

In 1946, after the war, I made the pilgrimage to the mecca of high energy physics, the University of California, Berkeley. At that time the Radiation Lab was the most exciting place in the world for high energy physics research.

E. O. Lawrence, inventor of the cyclotron and Director of the Radiation Lab had already received the Nobel prize in physics (1939). Ed McMillan, my dissertation advisor, and Glen Seaborg received theirs in 1951.

Other members of the faculty, while I was a student, Emilio Segre (1959), Owen Chamberlain (1959), and Louis Alvarez (1968) later received Nobel Prizes. My Ph.D. was awarded in 1951.

I spent the next 10 years at the Lawrence Berkeley and Livermore Labs carrying out high energy experiments at the bevatron in Berkeley and diagnostic measurements at weapons tests in Nevada and the Pacific.

Then the space age presented new opportunities for experiments at Aerospace Corporation where I studied protons in the radiation belts of the earth and the decay of the neutrons that produce them.

In 1967 I accepted a position at the University of California, Riverside as a professor of physics and director of the Institute of Geophysics and Planetary Physics.

For the next 25 years I taught physics courses and continued my research in space physics, entered the new field of gamma ray astronomy and became more involved in astrophysics. I changed to emeritus status four years ago and continue my interest in research and science education while living in Santa Barbara.

When teaching astronomy to non-science undergraduate students, it was apparent that many of the better high school graduates, those qualified for the University of California, had an inadequate understanding of science.

This concern has been amplified by the responses of students and the public to questions in the Gallup and other polls over the last 30 years.

In this column I will discuss science issues that are important in public policy. I plan to present material useful to people who want evidence for their views. I hope the columns will help citizens make responsible judgments on Federal, State and Local questions.

I will write about the scientific frontiers and recent discoveries. And I would like to assist in discriminating between science and pseudoscience.

To twist a one-liner usually reserved for teachers. Those who can, do. Those who can't, write. Hope you find the columns interesting.

2. Science and the Public

Polls of the public suggest a disappointing lack of knowledge about science. Too many people still believe in the paranormal and in pseudoscience. There even seems to be an escalating rebellion against science.

"How long does it take for the Earth to go around the Sun: one day, one month, or one year?"

A recent science survey by the National Science Foundation asked 2006 American adults that question. Only 47 percent gave the correct answer, one year. The result is not a lot better than the 33 percent expected by chance.

Maybe astronomy is not a fair subject--not even for the astrology buffs. How about biology? "The earliest human beings lived at the same time as the dinosaurs. (True or False)." Here 48 percent were correct with the answer, False--no better than the 50 percent expected from chance.

Try physics. "Electrons are smaller than atoms. True or False." The correct answer, True, was given by only 44 percent. Should we be worried? Does this result, even less than the 50 percent expected by chance, mean that the public has actually been misinformed?

Only 27 percent of the public answered seven of the 10 questions correctly, considered a passing grade. Perhaps this is not too surprising as the results of a recent international science competition of high school seniors from 14 nations found those from the United States were dead last.

What do students actually believe? Polls show that their pseudoscientific views have not changed much over the last 30 years. The pseudoscientific ignore the basic requirements of science, testable hypotheses and relevant evidence. Typical is the study [4] by Francis B. Harrold and Raymond A. Eve, faculty in anthropology and in sociology at the University of Texas, Arlington. The voluntary, anonymous questionnaire was given to 409 students at the University of Texas.

They found that 22 percent believe that aliens have visited the earth in the past--expected from students watching 'scientific' programs on TV talk shows. Not expected was the 41 percent that accept pre-Viking trans-Atlantic voyages. About as many believe in creation as in evolution for the origin of life. A whopping 65 percent trust in Noah's flood while a smaller 24 percent that creation occurred in six 24-hour days.

The paranormal beliefs--the Loch Ness Monster, Bigfoot-Sasquatch, UFO's as spacecraft, the Bermuda Triangle and Reincarnation were quite similar at a little less than 30 percent. Beliefs in "communication with the dead" and "ghosts exist" were 38 and 35 percent, respectively. The biggest surprises were the high 59 percent that believe "some can predict the future by psychic power" and the low 8 percent that "astrology predicts personality."

The belief of the American public in the paranormal is widespread. A poll [5] of 1,236 adults by George H. Gallup, Jr., and Frank Newport found that 27 percent of adults believe that extraterrestrial beings have visited the Earth at some time in the past and that 25 percent accept the myths of astrology.

They found that extrasensory perception, the ability to acquire and transmit information and predict the future without use of the 5 senses, sight, sound, touch, taste and smell, is believed by 49 percent of adults. They also discovered that 36 percent accept telepathy, the ability to communicate between minds without the 5 senses. And 26 percent accept clairvoyance, the power of the mind to know the past and predict the future.

The authors determined that 25 percent of the respondents believed in ghosts, that spirits of dead people can come back in certain places and situations; 21 percent in reincarnation, the rebirth of the soul in a new body after death; and 18 percent that people can hear from or communicate mentally with someone who has died.

The reasons for these beliefs in psuedoscience and the paranormal seem to result from poor or no teaching of science in the home and at school, and the lack of opportunity to learn about science on TV and in the newspapers. One-third of the students graduate from high school with only one year of math and more than half with only one year of science. Less than 20 percent of our high schools teach physics.

Above all, students should have the opportunity to learn in the lab--to experience the thrill of testing Newton's laws of force, observing the planets through an 8 inch telescope, preparing a salt from an acid and a base and watching the movements of tiny microbes under a microscope lens. Then experience the satisfaction that accompanies the verification of a theory by calculations carried out on a desk-top computer.

3. The Metric System

The metric system is used by all industrialized nations except the United States. Scientists world-wide calculate with it. Based on factors of ten, it is easier, faster and less prone to error than the English system. Yet Americans appear reluctant to make the change.

If Johnny doesn't like arithmetic, we blame it on his lack of effort, on the teacher, or the parents. However, the problem may just be our English system of measurements.

When cars made in Germany and Japan have less defects than those produced in the United States it could be due to the meter, kilogram, second--mks--metric system of measurements.

Our forefathers were brilliant in establishing a decimal monetary system based on powers of ten. We have pennies, dimes and dollars. Fortunes are measured in millions and billions of dollars and our national debt in trillions.

But they were not so bright when it comes to weights and measures. In the English system adopted, we must remember that 12 inches equals one foot, 3 feet equal one yard, 5.5 yards equal one rod and 320 rods equal one mile.

In weights (or mass), 16 ounces equal one pound and 2000 pounds equal one ton. For liquid volumes, one cup equals 8 ounces, one pint equals 2 cups, one quart equals 2 pints and one gallon equals 4 quarts, not to mention the units of teaspoons and barrels.

The length of a two by four is measured with a yard stick to be 3 yards 2 feet 5 1/4 inches in length. It is 3.813 yards, 11.44 feet or 137.3 inches long. In the metric system it is simply 3.487 meters, 348.7 centimeters or 3,487 millimeters long, the units differing by factors of 10s.

Years ago architects, engineers, and machinists replaced fractions of an inch or foot like 1/4, or 1/64 by their decimal equivalents.

Scientists have long worked in the metric system. In 1960 the International General Conference on Weights and Measures redefined the meter and second and switched to the International System using mks units.

The metric system was adopted because it is faster and easier to use and there is less chance for error. Even if the initial data are given in English units, scientists convert to metric, carry out the calculations and, if required, switch answers back to the English system.

The American driver can quickly learn to read signs and judge distances in kilometers and speeds in kilometers per hour. One mile is 1.6 kilometers so 65 miles per hour is 104 kilometers per hour. After all, we already run 5, 10 and 15 K races.

Height and weight will be in meters and kilos. The system of measurement of the articles we buy is similar to the decimal system of dollars we pay. Meat purchased at the supermarket will be in kilograms (kilos) and milk in liters.

Collar sizes will be in centimeters (2.54 centimeters equal one inch). A jacket size will be 102 centimeters instead of 40 inches and a trouser waist 81 centimeters in place of 32 inches.

Fortunately, our unit of time, the second, is the same used in the mks system.

Surveyors and carpenters can save time and effort by working in the metric system. And communication between the public and scientists is improved by using the universal mks language.

The metric system phase-in was started back in the 1970s. By Congressional mandate, Federal Agencies were required to switch to the metric system by 1992. However, conversion slowed when the Reagan Administration reduced funds for the change and eliminated the oversight National Metric Board. Now hardware and auto parts stores must carry duplicate screws, bolts, wrenches and other tools to supply both the English and metric markets.

The United States is the only industrialized country in the world whose production is not based on the metric system. Customers buy to metric specifications. American industry will not compete successfully in the world market unless it adopts the mks system of units.

4. Santa Barbara Museum of Natural History

The Santa Barbara Museum of Natural History, founded in 1916, has a unique collection of coastal birds, fish and mammals. Its museum, planetarium, demonstrations, tours, talks, and workshops make a real difference in the lives of children and adults in Santa Barbara and surrounding areas.

Nestled behind the Santa Barbara Mission, hidden in the California oaks in historical Spanish style buildings, is a scientific treasure, the Santa Barbara Museum of Natural History.

A 72-foot blue whale skeleton near the main entrance gives a dramatic introduction to life in the Santa Barbara Channel. Inside the museum, the visitor sees extensive exhibits of Santa Barbara-area plants and animals and their environments.

In the marine life hall, a 225-pound squid hangs overhead. Exhibits of sea stars, shells, sponges, sharks and rays greet the guests. Spectacular dioramas of a beach with tidepool, shells and birds; life beneath a pier; an intertidal zone; and a rocky intertidal habitat entertain and educate.

Birds usually seen in motion at a distance are skillfully arranged in the bird diversity hall with views close up for comparisons of their markings and other characteristics. To help later identification of Western Gulls at the beach, one exhibit shows the progression of markings from dark-spotted one-year-old young to the white breasted gray-winged four year old adults.

Impressive too are the exhibits of the large birds here and in the bird habitats hall. The brown pelicans that sail so gracefully in formation on the updrafts along the cliffs above the beaches and the blue herons, great egrets and cormorants are seen in their natural environments.

The display of the giant California condor is a reminder that the museum is active in the release and protection of the condors in the wilds of Santa Barbara County.

Children have fun in the insectary with a hands on microscope that gives close-up views of butterflies and insects. Butterflies mechanically flap their wings and bees are watched as they fly in the hive. The buzzing of bees and the chirping of crickets are heard at the push of a button.

Other exhibits--all musts--are the Chumash Indian life, fossils and geology and the astronomy center with its planetarium.

The museum was founded in 1916 as a Museum of Comparative Oology (study of eggs) but has progressed in 80 years into a sophisticated museum with first class research, education and public service.

More than 20,000 children with their teachers visit the museum every year and more than 14,000 attend docent-led tours, talks and planetarium programs.

Educational science materials are loaned to hundreds of school children. Many workshops and presentations are made each year for teachers and docents.

Over 20,000 people each year attend the planetarium programs at the Astronomy Center. The Sea Center, dedicated to sealife, was established in 1986 on Stearn's Wharf. They receive over 70,000 visitors each year.

Perhaps less is known by the public about the special research carried out by the professional curators of the museum. The collections and research division includes three departments--anthropology, invertebrate and vertebrate zoology.

The anthropology department has over 50,000 artifacts from western North America. Research and publications have emphasized Chumash archaeology sites, healing, mythology, environmental management and ethnohistoric studies.

In collaboration with Chumash research assistants and others, a computer data base has been organized containing baptismal, marriage, burial and genealogical data on more than 20,000 Indians.

The invertebrate zoology department has one of the ten most important mollusk (soft body animals such as octopuses, snails or clams) collections in North America. They have the historically significant Eyerdam collection, mostly land snails, and the Ferussac collection of 400 folio-size, hand colored engravings and lithographs of mollusks, dating from 1820-1851, among the finest ever produced. Their research and publications concentrate on octopuses, bivalves (clams) and land snails.

The vertebrate zoology department has an extensive bird egg collection considered among the country's ten most important. Studies of gray whale migration in the western Santa Barbara Channel include more than 1,800 sightings since 1975.

The museum also contains a library with over 40,000 volumes, open to the public, and a collection of the largest amount of island mammoth material in the world.

The new director, Robert Breunig, plans to continue and enhance the excellence in research, education and public service and strengthen ties with the nearby teaching and research institutions such as UC Santa Barbara. He hopes the Museum will make a difference in the lives of people in Santa Barbara.

5. THE SMITHSONIAN MUSEUM

Many members of the American Chemical Society, the American Physical Society and other scientists were disturbed by the treatment of science in an exhibition in the National Museum of American History. The American Chemical Society had contributed $5.4 million toward the exhibition.

The Smithsonian Institution is again taking heat for its exhibition of "Science in American Life," now showing in the National Museum of American History.

It was only one year ago that Martin Harwit, Director of the National Air and Space Museum, resigned. He was responsible for the planned exhibition, "The Last Act: The Atomic Bomb and the End of World War II."

American World War II veterans and many in Congress objected to the implications that the bombings of Hiroshima and Nagasaki were not for valid military reasons but rather for vengeance against Japan.

The American Chemical Society put up $5.4 million to sponsor the science exhibition but had no control over its content. Society chair, Paul Walter, in a February 1995 letter to Michael Heyman, secretary of the Smithsonian, complained that the last four years of planning were quite frustrating. The ACS advisory board was continually ignored by the Smithsonian staff and rebuffed, at times, by some arrogant project personnel.

The American Physical Society says the science exhibition exaggerates science failures and trivializes is accomplishments.

The exhibition omitted the very important space program and the invention of the transistor which led to computer chips with their vast applications. And it blamed science for environmental and other social problems.

The exhibit played up conflicts--between basic research and commercial goals, attitudes toward the atomic bomb, and stimuli for development of the birth control pill.

Social historians seem to look at science history differently than the scientists. Progress in science, they say, is influenced primarily by the social, political and economic environment of the time. They search for discords among ideas, groups and individuals. It is the conflicts along the way that count.

While conceding society does have some influence, many scientists insist that the real driving force for new discoveries is the scientific capability and technology of the times. Einstein was a genius but needed the Michelson-Morley experiment before proposing his Special Theory of Relativity. And the invention of high-power rockets was required for the Apollo Manned Space Program.

Scientific advancements in the 20th century have revolutionized our lives. New materials, such as plastics, were invented by scientists and engineers. They have developed new kinds of transportation: autos, jet planes and satellites. New means of communication--by telephone, radio, movies, television and computer--were introduced.

Discoveries of the properties and interactions of molecules, genes and DNA are changing our understanding of disease and its treatment, and dramatically affecting other fields such as anthropology and forensics.

Materials, conveniences and entertainment in our homes have reformed our way of life. The quality and quantity of food; and treatment of disease with new chemicals and radiations have enhanced and extended our lives.

The scientist's viewpoint on exhibits is not promotional and self-congratulatory as claimed by some curators. Nor do the scientists insist that science history be portrayed heroically. Indeed, science at times has been misused.

Unfortunately, the public has little knowledge or understanding of science. Many had slight contact with science in the schools. Much of the science they do get from television and the newspapers is negative, coming from environmental, animal and other activists. The views of a few cowboy scientists is given the same credibility in the media as the other 95 percent of mainstream scientists.

A science exhibition is one of the few places children and adults can learn about science. So scientists just ask that science exhibitions give science a fair shake, that the successes be emphasized as well as the warts. And that museum curators be careful about reconstruction and revision of science.

Fortunately, I. Michael Heyman, secretary of the Smithsonian, is now consulting widely with scientists as well as historians and curators about the future of the "Science in American Life" exhibition. And Admiral Donald Engen, retired, has recently been appointed (June, 1996) as the new Director of the Air and Space Museum.

6. Higher Superstition

Science is under attack from the 'academic left' as a 'construction' that is dependent on the social and political characteristics of the times. They claim there are other ways of knowing than the scientific method--not the standard view that science is knowledge that can be tested.

"Higher Superstition" [1] by life scientist Paul Gross and mathematician Norman Levitt raises important questions about the growing criticism of science by some humanists and social scientists.

This book is 'must reading' for university and college administrators, and equally important for alerting the public, concerned about education, to attacks on science from what the authors call the 'academic left'.

While alarming, the critics are still small groups but include highly visible faculty from some of the country's most prestigious universities. The movements include postmodernism, feminism, radical environmentalism and multiculturalism.

The hostility extends beyond scientists, to their institutional social structures, and education and professional training. The academic left disapproves of the content of scientific knowledge and of the idea that scientific results are usually reliable and based on reasonable methods.

The authors point out that 'postmodernism' is lodged mostly in the fields of literary criticism, social history and cultural studies. Strong 'cultural constructivism' maintains that science is a highly elaborate set of conventions of our particular culture--not the standard view that science is a body of knowledge that can be tested by experiment and observation.

The radical feminists view science as poisoned and corrupted by 'patriarchal' gender bias. They claim suppressed female scientists have been unable to penetrate 'official science'. Thus the world is deprived of alternative points of view, of women's ways of knowing. However, no chemistry or physics woman faculty member I know, is actively campaigning for courses in 'woman's quantum mechanics'.

We are all concerned about our environment, preserving plants and animals, clean air and water. Jeremy Rifkin in "Beyond Beef" [6] laments, "The modern era has been characterized by a relentless assault on the earth's ecosystems. Dams, canals, railroad beds, and more recently highways have cut deeply into the surface of the earth, severing vital ecological arteries and rerouting nature's flora and fauna. Petrochemicals have poisoned the interior of nature, seeping into animals and plants, soaking the organs and tissue with the tar of the carboniferous era." But hysterical declarations of radical environmentalists do not solve the problems. Scientific research and technology do.

A common bond among the movements seems to be their fixation on the 'uncertainty principle' in quantum mechanics and 'chaos' in dynamical systems theory. They interpret these developments as removing predictability from physics laws and permitting alternative possibilities. However, these theories have actually improved prediction.

The uncertainty principle states that the product of the uncertainties in momentum and position of an object is approximately equal to a quantity called Planck's constant. This is important in small systems like atoms and nuclei where both the momentum and position of an electron cannot be determined precisely.

In everyday situations both quantities are so precise neither uncertainties are measurable. For example, suppose a baseball pitcher throws a fastball at a velocity of 95 miles per hour with an uncertainty in velocity at the plate of one part in a million billion. The uncertainty principle gives the uncertainty in position of the ball as one billionth of a billionth of an inch.

This extremely small uncertainty is no surprise as the laws of classical mechanics have been tested over and over at the distances and velocities encountered in our lives. Classical mechanics has not been eliminated, only subsumed by quantum mechanics that is needed for very small energies and short times.

It has been known since Newton's day that certain non-linear dynamical systems could grow extremely fast. Even with nearly identical initial conditions the systems would diverge and have widely different end results. Chaos theory offers possible ways to improve prediction in these complicated problems. It helps find answers to questions that have been intractable in the past.

The critics are still attacking only at the fringes of science. But the authors warn that the "long-term consequences of these trends--for science education and for public judgment of scientific issues--may be infinitely more serious than the 'political correctness' wars currently being waged on university campuses."

7. The Culture Wars

Lord Shaftesbury in "Letters to his Son" in 1774 said that ridicule is the best test of truth. Sokal's parody may be the beginning of the end of the 'social construction movement'.

"But the Emperor has nothing on at all!" cried a little child. As in the fairy tale by Hans Christian Anderson "The Emperor's New Clothes," the trendy academic discipline of "cultural criticism" has just been exposed.

A major journal for cultural studies, "Social Text," published unknowingly as a serious article, a spoof [7] by physicist Alan Sokal, a Professor at New York University. The same day he revealed the hoax in the magazine “Lingua Franca” [8]. The satire was Sokal's way of countering the misinformation and pretensions of the 'Social Construction Movement'.

A report of the prank appeared May 18 on the front page of the "New York Times." It was followed on May 23 by several letters to the editor.

The Social Constructionalists claim that science and scientific methods are only social constructs and that truth is subjective. Objective reality is fundamentally unknowable. Scientific knowledge is not a version of universal truth that is the same at all times and places.

They interpret the 'uncertainty principle' in quantum mechanics and 'chaos' in dynamical systems theory as examples of unpredictability in physics and mathematics. That is nonsense. In fact, both improve the ability to predict--from the smallest systems in atoms to the largest systems involving groups of galaxies.

Sokal's article in the "Social Text" was written as a normal contribution with the usual references and footnotes. He used the deconstructive jargon and buzzwords cherished by the editors and played to their prejudices. He says it wasn't difficult to write because he wasn't constrained by any standards of evidence or logic.

In an interview with the "Times," Andrew Ross, co-editor of "Social Text", said that about a half-dozen editors of the Journal dealt with Sokal's manuscript. They read it as "about the relationship between philosophy and physics." But no advice was requested from referees either in physics or philosophy. This was a grave mistake because they would have immediately spotted the parody.

A contribution [9] to the "Times" OP-ED page from Stanley Fish, editor of "Social Text" and Professor of English and Law at Duke University, labeled the trick a "bad joke." M. Kreitzer in a letter to the editor [10] says, "Mr. Fish has made a scholarly contribution to a complex and obtuse field, provoked by Mr. Sokal's prank. He (Sokal) has done us all a service."

Fish draws an example from baseball to make his point. He guides the reader to agree that balls and strikes are both socially constructed and real. The rules of the game have changed in the past and could change in the future. So balls and strikes are real but they can change. We agree. But the science undergirding the game will not change.

He ignores the science, the forces imparted to the ball by the pitcher, that cause velocity and spin when he throws; the effect of the air on the trajectory; and the pull of gravity that causes the ball to drop. These are real, do not depend on the President, Congress, the Baseball Leagues, are not socially constructed and will be the same in all places and times as long as the physical conditions are not changed.

Jerry Coyne, Professor of Ecology and Evolution at the U. of Chicago, in a letter to the editor [11] of the "Times", suggests that the cultural pundits "have their children vaccinated, take antibiotics when they are infected and ride in airplanes," but claim in their own work that scientific facts are as transient as "trends in culture or philosophy." And quotes George Orwell, "One has to belong to the intelligentsia to believe things like that: No ordinary man could be such a fool."

On the West coast, Ruth Rosen, Professor of History at UC Davis, assisted in the expose. In a column [12] in the "Los Angeles Times," she made the analogy of Sokal's satire to the Emperor's New Clothes.

She said that the academic scandal actually began about a decade ago. The Academic Emperors "focus obsessively on the linguistic and social construction of human consciousness, not on the hard reality of people's lives. Their claim to originality is particularly offensive to historians who have always known that social structure and cultural meaning change over time. With few exceptions, their pretensions obscure their nakedness."

8. Pathological Science

'Cold fusion' and 'remembering water' are just two of several examples that could be cited of 'pathological science'. Cold fusion was discredited by careful experiments of scientists from other labs in the year following the press release by Pons and Fleishmann. Remembering water of Benveniste and colleagues was discredited by a visiting team of investigators from "Nature." As with other pathological sciences both are slowly fading into obscurity.

Irving Langmuir, 1932 Nobel prize winner in chemistry defined "Pathological Science" as "the science of things that aren't so."

In those experiments, he says, the maximum effect of the causative agent is usually barely detectable above background, and the results are often erratic and essentially independent of the agent's dose or intensity.

Claims are often made of great accuracy and results and theories contradict other scientific knowledge. Criticisms are answered by improvised excuses. Support from other investigators quickly rises to a maximum, then gradually fades away disregarded.

Such is the case of "Cold Fusion" announced by electrochemists Stanley Pons and Martin Fleishmann at their famous press conference [16] at the University of Utah on March 23, 1989. The public was entranced by the promise of cheap, convenient, clean fusion energy at room temperature that would be available to every home, car, business and industry.

This flew in the face of 60 years of nuclear energy experiments and 40 years of controlled thermonuclear fusion research. Now, seven years later the press and the public have forgotten. A few of the faithful still meet once a year at an international conference on cold fusion. They are not taken seriously by the rest of the scientific world. Cold fusion has moved into oblivion.

In 1988 an article [13] appeared in "Nature" on 'remembering water' that astounded the scientific world. The report, by French biochemist Jacque Benveniste and 12 other biologists, 8 French, 2 from Israel and 1 each from Italy and Canada, was titled "Human Basophil Degranulation Triggered by very Dilute Antiserum against IgE."

After the antibody serum molecules were removed from the distilled water, the authors argue, the water still "remembers" the serum's chemical properties. The antiserum molecules somehow cause water molecules to rearrange so they act like the missing serum itself. And the more the dilution the stronger the effect.

Martin Gardner reported [14] in the "Skeptical Inquirer" that the dilution was so great not even one molecule of the antiserum was left in its solvent.

Because the editors of "Nature" considered the results unbelievable, they included an accompanying disclaimer.

They agreed to publication so that the authors could give an accurate report of their work, already widely discussed in the popular literature in France. The editors also wanted to provide information that would enable other scientists to confirm or falsify the exceptional claims. Nature then sent a visiting team to visit Benveniste's lab.

Although the authors' finding violates the fundamental laws of physics and chemistry it was seized by 'homeopathy' supporters as verification of their curative methods. Homeopathy is a medical practice that treats a patient by giving a minute dose of a remedy that produces symptoms of the disease in a healthy person.

In the 19th century, homeopathy prospered. In 1900, in the US, there were 14,000 practitioners and over 100 homeopathic hospitals. Homeopathy was taught in 20 schools.

By 1960 it had almost disappeared, but was revived in the '60s and '70s by the New Age followers, joining the fads of holistic medicine, acupuncture, herbal remedies and natural foods.

9. Science and Technology by Mandate

Battery electric cars will replace current ones only when the proper batteries are available. It does the switch to battery electric cars no favor to claim otherwise.

The California Air Resources Board (ARB) has learned the hard way that no governmental agency can mandate a new science discovery or technological break-through.

In March 1996, the ARB repealed the regulation that two percent of the autos for sale in 1998 be exhaust pollution free (electric battery powered). The increasing percentage through 2002 was eliminated but the 10 percent by 2003 was retained.

The reversal was not caused by stalling of the auto manufacturers nor by a conspiracy of oil producers. But the ARB was finally convinced that a battery capable of giving needed distance of travel, velocity, acceleration and convenience, at a cost the public would pay, would not be available before 2003, at the earliest.

Using batteries to power vehicles is not a new idea. Early this century Thomas Edison and other inventors spent years trying, without success, to develop a battery with high enough specific energy (energy divided by mass) to compete with the internal combustion engine. They failed. The need for such a battery has been clear since then, and chemists and engineers keep trying to develop the battery.

Mandate of science and technology by government just doesn't work.

No government official in 1811 commanded John Dalton to suggest that the molecules of a particular chemical compound are all alike, the start of the science of chemistry.

Petroleum molecules consist mostly of hydrogen and carbon. Energy is released when atoms of hydrogen or carbon combine with atoms of oxygen to form molecules of water or carbon dioxide.

Oil was available to the ancients as far back as 3000 BC. There was a need through the ages, as now, for improved methods of transportation. But petroleum could not power vehicles until the understanding of the laws of thermodynamics, of organic chemistry and the development of the internal combustion engine in the late 1800s. The combustion engine was not invented because of a government edict.

Chemical reactions can be controlled, or they can run away as explosions. In 1867, Alfred Nobel, founder of the Nobel prize awards, invented dynamite, then other explosives used in chemical weapons. Repetitive controlled small explosions of gasoline and oxygen in the cylinders of internal combustion engines propel cars. Neither was forced by the regulators.

Not the development of nuclear physics in the early 1900s, nor experiments leading to the discovery of nuclear fission were mandated by any civilian or government organization. Extensive studies of the nucleus of the atom by pioneering nuclear physicist, Ernest Rutherford, and others led to the discovery of nuclear fission of uranium by Otto Hahn and Fritz Strassmann in 1939. Energies per reaction, a million times those of chemical reactions, were emitted in the process.

Nuclear physicists suggested that controlled fission in reactors might lead to an inexpensive source of electrical power, and to energies released in weapons millions of times those in chemical bombs.

President Franklin D. Roosevelt did not order American scientists to develop nuclear fission reactors or bombs. Instead, the physicists shared their information with the government. Italian physicist Enrico Fermi and his coworkers established the first working reactor under the football stadium at Soldier's Field on the campus of the U. of Chicago, just two years later. And fission bombs were dropped on Hiroshima and Nagasaki, Japan, in 1945.

John von Neumann and co-workers designed the first digital computer in 1946. No government body required them to produce computers by 1996 that could process information with speeds greater than a billion bytes per second. They first used vacuum tubes. Later research replaced these by transistors.

Then came integrated circuits. Minicomputers appeared in the 1960s and microprocessors in 1971. Larger and larger memories have been manufactured on a single chip. And higher and higher densities of components giving shorter and shorter computation times are being produced.

There is a market for battery electric vehicles that give performance, and at the same time are pollution free. When the technology is ready, plenty of entrepreneurs will accept the challenge. The major manufacturers of cars with internal combustion engines will be first in line.

10. Forensic Use of DNA

It seems the public still has little confidence in the forensic use of DNA. The jury found O. J. Simpson innocent of the murders of Nicole Simpson and Ron Goldman in the most publicized trial of the 1990s. The jury apparently ignored or devalued the vast amount of DNA evidence presented by the prosecution.

DNA identification has been praised by police, politicians and much of the public as the greatest breakthrough in forensic science since fingerprints were introduced in 1892. They are enthusiastic about its qualifications for fixing guilt or innocence.

For more than a year, TV and the press were obsessed with the O. J. Simpson trial. DNA and restriction fragment length polymorphism--or RFLP--were household terms.

Forensic DNA was first used in England in 1986 and in the United States one year later. Although soundly based on molecular biology, early DNA typing was hurt by poor procedures, standards and interpretation.

A three year study by the National Research Council, released in 1992, addressed the problems and emphasized the importance of laboratory accreditation, rigorous quality control and external blind proficiency tests.

Eric Lander and Bruce Budowle were principals on opposite sides in the debate over forensic DNA typing. In their paper [15] in Nature, they agree that the scientific issues have been resolved. "(We) could identify no remaining problem that should prevent the full use of DNA evidence in any court....The DNA fingerprinting wars are over,"--except for population genetics which is still an issue.

In DNA typing, the pattern of measured lengths of selected DNA fragments, RFLP, from a sample collected at the scene of the crime is compared to the pattern from a sample of the suspect's blood. If they disagree, the suspect is cleared.

If the blood samples match, before incriminating the suspect, the probability of a match by chance, is addressed. This 'match probability' is the probability that a randomly chosen person from an appropriate population would match the crime DNA pattern.

To decrease the probability of a match by chance, several sites on the DNA are matched and the 'product rule' is used. The resultant probability of a chance match is the product of the individual site probabilities if they are all independent. It's like flipping a coin where the probability of four heads in a row is the product (1/2x1/2x1/2x1/2) = 1/16.

However, the probabilities are independent only for well mixed populations, not necessarily so for sub-populations (e.g., race or geographical location). The NRC committee settled this question by the conservative 'ceiling principle'. A strict upper bound to the product probability occurs if the maximum of each of the single possible sub-populations is chosen. This gives every benefit of doubt to the suspect.

Even with this ultra conservative approach, e.g., a 1/30 maximum chance for each sub-population for four DNA locations gives a probability of chance match of slightly more than one millionth and for six sites slightly more than one billionth.

This conservative estimate does not prohibit expert witnesses from giving, in addition, a best estimate of the probability based on more detailed information about the sub-populations that has an even lower probability of chance match.

In January, 1996, the California 2nd Appellate District Court approved the RFLP-DNA profile tests for use as scientific evidence. This decision is binding on all Superior and Municipal Court judges in Los Angeles, San Luis Obispo and Santa Barbara Counties.

The O. J. Simpson trial was not affected by this court ruling as Simpson's attorneys earlier waived their right to contest the admissibility of the RFLP tests.

Forensic use of DNA has also proved useful in providing the release of innocent prisoners. Sixteen years ago Kevin Lee Green was convicted of killing his unborn baby and of severely beating his wife.

He declared that he was innocent, that he went out for a hamburger and returned to find his wife unconscious. His wife testified that at first she remembered nothing, but later, after reading a baby magazine, recalled the attack by her husband.

Green was released in June, 1996 when the DNA of Gerald Parker, a convicted rapist in prison for parole violation, was connected with blood that had been saved from the scene of the crime.

11. Gambling

If there is any question about who gets rich from gambling, check out Las Vegas and Atlantic City. Las Vegas is adding a new billion dollar hotel and casino a year. Atlantic City has announced new casino and hotel projects worth more than $4 billion. That major growth industry of today is financed by your gambling losses.

The good news is that UC Irvine math Professor Mark Finkelstein has found a way to better the odds in the 'Super Lotto' jackpot. The bad news is that he must play for the next 2.3 million years to benefit.

In super lotto, six numbers are picked out of 51 that run from 1 to 51. In the first draw the chance of picking the number drawn is one in 51. In the second draw with 50 numbers left, the chance of picking the number drawn is 50 divided by 2, as it could have been picked in either of the first 2 draws.

Continuing through 6 draws gives 18,009,460 possible combinations--(51 x 50 x 49 x 48 x 47 x 46) divided by (1 x 2 x 3 x 4 x 5 x 6). The chance of picking those 6 numbers is then 1 in 18,009,460.

Suppose 20 million one dollar tickets are purchased. The pot is then $20 million. However, only 50 percent of the pot goes to the winners, 34 percent goes to public education and 16 percent to retailers and operating expenses.

If there is no winner the pot keeps growing with more players until there is a winner. On the average the lottery player loses 50 cents of every dollar spent. No money advisor would recommend lotto for an investment portfolio. The player, however, is making a significant contribution to education.

In games of pure chance, each play is independent of every other play. The future does not depend upon the past. Any one play can deviate from the odds but the odds are increasingly accurate as the number of plays increases.

A typical slot machine has 3 reels with 20 different symbols, including one 'bar' on each reel. If each reel is independent and random, the chance of obtaining 3 bars is 1/20 x 1/20 x 1/20, or 1 chance in 8000. So the house can stay in business, build expensive hotels and casinos and pay a profit to the owners, the payoff to the player is reduced from the correct odds.

The payoff withheld varies from about 5 to 15 percent, depending on the casino and location of the 'slots' in the casino. Because of the many plays by most bettors, the experience more closely approaches the calculated laws of chance. Slots are not likely to provide a living.

In European casinos, roulette is king. A wheel with 38 pockets in the United States, 37 in Europe, is spun. A ball, by chance, stops in one of the pockets. The pockets are numbered from 1 to 36 with 0 added in Europe and 0 and 00 in the America. The player wins if the ball falls on his number. The house takes 0 and 00. No wonder it's more popular in Europe where the payoff is 0.973 compared to 0.947 in the United States. Other bets are also possible.

Two dice are rolled in the game of craps. Each of the six sides of one die can be matched with any of the 6 sides of the other. Each die is independent so the chance of throwing a 2 or 12 is one-sixth x one-sixth, or 1 divided by 36. The odds is the ratio of the unfavorable to favorable outcomes, 35 to 1.

A 7 may be rolled by 1 and 6, 2 and 5, 3 and 4, 4 and 3, 5 and 2 and 6 and 1--6 ways out of 36 possibilities. The chance is then 6 in 36 or 1 in 6 and the odds 5 to 1. In this case the payback may be reduced to 4 to l. This amounts to withholding one-sixth or 16.7 percent of the wager. For other combinations the odds are calculated in similar ways.

There are many different types of wagers in craps with the house returning from 0.6 to 27 percent less than the correct odds. The sophisticated gambler knows that his bankroll will last longer if he makes bets close to the 0.6 house break.

Years ago when black jack was played with one deck of cards, bettors could remember the cards played and use them to beat the payoff odds for normal play. No longer. The casinos responded with multiple decks, frequent shuffles, and other means to stop the winnings.

The gambler's bankroll fluctuates up and down. Most bettors know their limitations, set maxima for their losses and stop gambling when that limit is reached. Bettors seldom stop playing as winners, almost always as losers near the maximum they can afford to lose. This is an additional bonus for the house. The quitting bankroll is no longer distributed as losses around the average house take, but rather about the player's maximum acceptable loss.

The most tragic cases are those of gamblers who play beyond their means, and continue betting with borrowed money. The casinos make this easy by accepting credit cards to maximum credit and personal notes impossible to repay. Casinos are not in the business to lose money.

CHAPTER III

MEDICAL SCIENCE

The 20th century has seen an exploding accumulation of medical knowledge by science, and its application to human health and the fight against disease. In the 95 years from 1900 to 1995 the life expectancy of Americans increased from 52 to 80 years for women and from 48 to 73 for men. This increase of more than 50% in less than a century is phenomenal when compared to the slow increase in life expectancy by a factor of two in the preceding 2000 years.

BRIEF HISTORY OF MEDICINE

Medicine before the 18th century was mostly folk medicine carried out by medicine men who relied on mysticism, magic and religion [1]. The ancients treated common illnesses, like colds, with their favorite herbs. Serious diseases were caused by spells cast by angry gods, enemies and intruding objects. To cure the patient it was necessary to cast out the demons. Holes were drilled in skulls to permit the intruders to escape. Medicine bags and charms were worn to ward off evil spirits and bring good luck.

Ancient Hindus in India believed that the elementary materials, air, phlegm and bile, must be balanced for good health. Nutrition was prescribed. Hundreds of medicinal plants and minerals were listed. Diseases were combated with emetics (vomit inducers), purgatives (cleansers) and enemas.

Leeching, cupping (suction) and bleeding were common practice. Surprisingly, these are still used today by some folk healers in a few parts of the world. Even in those days surgery drained abscesses, removed tumors and amputated limbs.

Medicine in China dates back to Emperor Huang Ti in the third millennium BC. The traditional Chinese medicine (TCM) is still popular today [2]. According to TCM all illnesses are caused by an imbalance of vital energies in the body. These energies, Qi or Chi, permeate everything in the universe. Variants of Qi energy, Yin and Yang, stream through meridians that are invisible channels in the body. Advocates claim therapies such as acupuncture, breathing, exercises, massage and moxibustion; herb medicines; and particular foods balance Qi and restore or maintain health.

Acupuncture therapists insert and twist flexible metal needles into the skin and tissue. Followers believe that the needles affect the distribution of the yin and yang in the meridians. In moxibustion, a small cone of moistened leaves is placed on the skin, ignited and crushed to make a blister.

Ancient Greek medicine established temples for the ill that were combination hospital and health resort. Baths, diets and exercises played important recovery roles. Gardens, fountains, theaters and stadiums holding athletic contests added to the rehabilitation. After incubation (temple sleep) the patients returned home cured.

The Greek philosophers, Pythagoras and others, led the transition from mysticism to reason in an attempt to understand nature. By 400 BC, Hippocrates established himself as the 'father of medicine'. We are familiar with the 'Hippocratic oath' that is still the ethical code of modern physicians. Aristotle, a student of Plato and teacher of Alexander the Great, is considered the first great biologist. His ideas dominated scholars until the renaissance about 1300 AD. The center of learning shifted to Alexandria in 300 BC with its famous library and medical school.

During the Middle Ages, medicine was preserved and extended by the Christians, Jews and Muslims. The Muslims established a renowned school and hospital at Jundi Shahpur in southwest Persia. Some of the compounds made by the early alchemists were later found to be useful as medicines.

The earliest medical school in Europe was established at Salerno in southern Italy in the 12th century AD. Other schools quickly followed. Medieval physicians examined patients' symptoms and diagnosed illnesses.

They prescribed baths, diet, exercise and rest; or emetics, purgatives and bleeding. Surgeons repaired fractures and hernias, and carried out amputations. They prescribed opium and alcohol for easing the pain. Midwives assisted women at childbirth.

Medical science developed slowly. Andreas Vesalius, a Professor of Anatomy at the University of Padua in Italy, published his seminal book "On the Structure of the Human Body" in 1543. William Harvey, a great experimental anatomist, developed his theory of the circulation of blood in 1628. He then founded the science of embryology (the formation and development of embryos) in 1651. The microscope was invented about 1600 and immediately applied to accurate measurements of bacteria and other microscopic size objects.

During the 1700s surgery and obstetrics became respectable sciences. The stethoscope was invented and became an indispensable tool for listening to the heart and lungs. The diagnostic practice of listening to the sound from chest tapping is still used today. Vaccination by inoculation with material from cowpox was found to protect against smallpox and virtually eliminated the disease. Fresh fruit and citrus juice were found to prevent scurvy.

Two pseudoscientific doctrines, Mesmerism (belief in a mysterious force that enables an animal or person to hypnotize others), named after Franz Mesmer; and Phrenology (belief that contours of the skull measure mental abilities and character traits) were popular throughout the 19th century. More recently they have been rejected as unsupported by scientific evidence.

The 19th century marked a new era of scientific medicine with increased progress in chemistry, physics and the biological sciences. The importance of the cell as the basic unit of living things was discovered by Rudolf Virchow, a prominent medical scientist.

The cause of infections in many diseases and in surgery was found to be microscopic living organisms we call germs. Louis Pasteur, a brilliant French chemist, established the science of bacteriology by a pioneering series of experiments. He demonstrated that fermentation of wine and souring of milk were caused by germs. He was responsible for, and gave his name to the pasteurization of milk. He is also noted for inoculations to prevent anthrax in cattle and sheep and rabies in dogs and humans. By 1900 tuberculosis, cholera and many other microorganisms were identified.

Joseph Lister revolutionized surgery by placing antiseptic carbolic acid between the wound and germs in the atmosphere. His name is preserved by the disinfectant Listerine. Disinfectants for hands and clothes and sterilization of surgical instruments became standard medical practice.

The general anesthesia agents, nitrous oxide and ether, then later chloroform were developed for dulling pain throughout the body during surgery.

Before the end of the 19th century, it was found that malaria and yellow fever were carried and spread to humans by mosquitoes. By controlling them, the United States was able to complete the Panama Canal by 1914.

The climax to the century of progress in medical science, was the unexpected discovery of x-rays by Wilhelm Roentgen in 1895 and radioactivity of radium by Pierre and Marie Curie in 1898. During the 20th century, these have proven important tools for diagnosis and treatment of diseases, especially cancer of the internal organs.

In the 1900s, the contributions of medical science to world health continued its rapid acceleration. Landmark discoveries in biochemistry, cell biology and physiology brought new understanding of the basic causes of disease.

New instruments and techniques from physics such as computerized axial tomography (CAT) scans with x-rays or gamma-rays: nuclear magnetic resonance (NMR) using spin properties of nuclei; and ultrasound (sonar) waves provided detailed images of internal organs. High energy accelerators provided 10s of MeV (million electron volt) energies of gamma rays, neutrons and charged particles for bombarding and destroying cancer.

Electron microscopes gave detailed pictures of the smallest viruses down to molecular size. Mumps, measles, German measles and polio viruses were identified.

Paul Ehrlich in Germany introduced the chemotherapeutic era in medicine with the chemical arsphenamine, sold as Salvarsan. It and its derivatives were responsible for bringing syphilis under control. In 1932, Gerhard Domagk in Germany discovered that the red dye Prontosil would attack streptococcal infections. The antibacterial agent, sulfanilamide and other sulfonamides with even greater potency received wide medical acceptance.

The most famous drug of the first half of the 20th century was penicillin. The antibiotic was discovered by Alexander Fleming, in England, when he observed that a plate culture of staphylococcus bacteria was inhibited by a chance mold, later identified as penicillin. Sufficient quantities were produced to be of significant value in World War II.

Penicillin was not active against tuberculosis (TB), but streptomycin, derived from cultures of a soil organism, was very effective. Later, when TB became resistant to streptomycin, other drugs such as aminosalicylic acid and isoniazid were satisfactory replacements.

The science of immunology was first developed to combat bacterial diseases. A vaccine for typhoid was produced from killed typhoid bacilli that prevented typhoid fever. In World War I, tetanus was brought under control by injecting into patients a sterile solution of antibody globulins (blood proteins) from immunized horses and cattle. Later in the 1930s an effective vaccine was developed against tetanus by injecting a toxin produced by the microorganisms. The immunity is produced by the stimulation of the body's own immune system. Effective vaccines were also developed in similar ways for diphtheria and whooping cough.

Vaccines for viruses, except for smallpox, were not developed until the 1930s when the technique for growing viruses in tissue cultures was perfected and the electron microscope invented. Vaccines for yellow fever, then influenza and finally poliomyelitis, by Jonas Salk, were produced. In 1960, the oral vaccine introduced by virologist, Albert Sabin, received wide use. Vaccines are now available also for common and German measles.

English physicist Francis Crick and American geneticist James Watson, both at the Cavendish Laboratory at Cambridge University, determined the structure of deoxyribonucleic acid (DNA) in 1953. This set the stage for the study of the control by DNA of the body's immunization system. DNA's role has dominated endocrinology (science of the endocrine glands), genetics (science of the genes) and studies of cancer and other diseases in medicine in the last half of the 20th century.

One of the successes of the science of endocrinology has been the understanding of diabetes--the disease caused by a failure of the pancreas to secrete insulin that controls the metabolism of carbohydrates. Since insulin was discovered in 1921, its injection has made it possible for a diabetes patient to live a normal healthy life.

Another success was the observation that rheumatoid arthritis can be eased by cortisone from the adrenal gland. Cortisone and its derivatives are strong anti-inflammatory agents. And an understanding of sex hormones led to new methods of contraception. With the pill it is possible to prevent ovulation and thus pregnancy. Hormones are also prescribed for many other medical problems.

For those concerned about nutrition, the discovery of the need for vitamins in the diet, in addition to carbohydrate, fat and protein, was a very important advance. Rickets is caused by a deficiency of vitamin D, scurvy by vitamin C (ascorbic acid), beriberi by vitamin B1 (thiamine) and pernicious anemia by vitamin B12. Vitamin supplements are more popular in the United States today than ever before with sales in the billions of dollars.

Plenty of problems still remain. Heart disease causes 33 percent of the deaths followed by cancer with 24, cerebrovascular disease (brain and blood vessels supplying the brain) with 7, and accidents with 4 percent.

The understanding of the cause of cancer, its cure and possible vaccination drives much of cancer research. While no cure is known, surgery to remove malignant growths is successful in many cases. Radiation treatments with x-rays and gamma-rays from cobalt radioactive sources are often able to retard or stop their growth. Chemotherapy has been useful in treating leukemia and other forms of cancer, often slowing their spread.

Acquired immunodeficiency syndrome is spreading rapidly in the United States and the rest of the world. AIDS patients have insufficient antibodies to fight life-threatening diseases like pneumonia and eventually die. The evidence is strong that patients are first infected by the retrovirus HIV. It then inserts its own genetic material into that of its host thereby causing AIDS. A large body of medical science is searching for a cure.

At the July, 1996 international meeting on AIDS in Vancouver, British Columbia, cautious hope was expressed by some scientists. Combinations of drugs in a few trials have slowed down AIDS. However, the trials have run for only a year and it sometimes takes up to a decade for the HIV infection to progress to AIDS.

ALTERNATIVE MEDICINE

Alternative medicine is the umbrella covering many folk remedies outside the mainstream of medical science. Many originated in ancient times and have not changed significantly over the centuries. They have their origin in magic, mysticism and religion and are not supported by scientific evidence. Advocates defend their benefits to society by anecdotes of cures rather than scientific studies.

Medical scientists have found that alternative treatments supported by enthusiastic therapists and receptive patients often succeed for reasons not controlled--instead of by the therapy. Therefore, it is very important to carry out repeatable controlled tests. Usually double blind tests are required--tests in which neither the subject nor the experimenter knows who receives the experimental dose and who receives the placebo.

One of the oldest and most popular alternatives is Traditional Chinese Medicine (TCM). It makes the basic claim that the vital energy Qi flows along meridians through the body. The goal of TCM is to balance Qi and to maintain or restore health. However, that Qi has never been identified and measured.

Nineteenth and 20th century physics has identified the forces and energies of nature including those in the human body. Extremely sensitive instruments have measured electric currents and electric and magnetic fields. But no energy has been identified with the properties attributed to the vital energy Qi. The lack of evidence for Qi, leaves TMC without scientific support. With no bonding theory, each of the therapies--acupuncture, moxibustion and herb medicines--must be justified on its own.

This does not mean that TCM therapies such as deep breathing, exercise, massage and diet are not beneficial to health. They are. These are all practices recommended by conventional medicine supported by scientific tests and theories of medical science.

Acupuncture, moxibustion and herb medicines may also have positive health benefits. They certainly have many advocates and are supported by lots of anecdotes. But before they can be accepted as beneficial medical treatments it is necessary that they be subjected to the tests of science. As with conventional medicine they must pass the rigorous double blind tests or other tests to obtain evidence for their validations.

The sale of food supplements, including herbs, is now on the rise. At the urging of Senator Orrin Hatch of Utah, Congress, in 1994, passed the Dietary Supplement Health and Education Act that changed some of the regulatory powers of the Food and Drug Administration. It removed restrictions on most claims made by food processors, as long as they do not appear on the food labels.

Some people think that herbs at worst are harmless and can only do damage to the pocketbook. But that is not true. Because a food or herb comes from plants and is called 'natural' does not make it safe to take internally. The herbal preparation, ma huang, used extensively in dietary supplements and mood altering drugs, has caused deaths to consumers. We all are aware that certain mushrooms are poisonous. Deaths have also been caused by coffee enemas claimed to retard cancer and other diseases by detoxification.

Senator Tom Harkin in 1991, then Chairman of the Senate Appropriations Committee and a user of alternative medicine, was also responsible for the mandate by Congress that set up the Office of Alternative Medicine (OAM) at the National Institutes of Health (NIH).

One observer on Capitol Hill said devising a plan to study unconventional medicine was like "orchestrating a roomful of cats....or setting the agenda for a convention of anarchists" [3]. Each of the many alternative therapies distrust the "medical establishment." Few are skilled at taking data or have the background and experience to direct clinical research.

Dr. Joseph Jacobs, the physician who became director of OAM in 1992, abruptly resigned two years later. "Science" said he "blasted politicians--especially Senator Tom Harkin,.....and some advocates of alternative medicine for pressuring his office, promoting certain therapies, and.....attempting an end run around objective science" [3].

In its first year, before Jacobs arrived, OAM funded "antineoplaston" (anti-cancer) therapy devised by Stanislaw Buraynski, a Houston physician. OAM paid the Burazynski Institute to prepare samples of amino acid derivatives to be tested by clinicians at NCI, the Mayo Clinic and the Memorial Sloan-Kettering Cancer Clinic on adults with recurrent brain tumors. During the first year the clinics tested only 3 patients.

Gina Kolata in the New York Times details [4] some of the funding by AOM for alternative medicine. In the second year, exploratory grants of $30,000 each were given to 30 therapists to "identify promising areas of future research." Among the awards given, after peer review, were acupuncture for depression, massage to stimulate AIDS patients' immune systems, hypnosis to speed bone healing, music therapy for patients with brain injury, dance for cystic fibrosis, macrobiotic (chiefly whole grains) diet to control cancer, yoga to control heroin addiction and prayer to control drug abuse.

Dr. Wayne B. Jonas replaced Jacobs as director of AOM. He had a family practice that included alternative medicine. He has been trained in homeopathy, bioenergy therapy, diet and nutritional therapy, mind/body methods, spiritual healing, electro-acupuncture diagnostics and clinical pastoral education.

The AOM is now funding ten research centers at about one million dollars each, for a total of $9,744,535. The first million dollar grant went to Ann Taylor, Professor of Nursing at the University of Virginia, Charlottesville to use magnets to relieve chronic back, foot and other body pains. This proposal is not supported by any scientific theory of the cause of pain or why the interaction of magnetic fields with the body should reduce (or increase) pain or by any scientific experimental evidence.

Another went to Dr. Leanna Standish, a naturopath at Bastyr University in Seattle that trains naturopaths. Naturopathy is a therapy that avoids drugs and surgery and emphasizes the use of 'natural' agents such as air, water and sunshine and emphasizes physical means such as manipulation and electrical treatments. Instead of the usual scientific tests, she plans to study about 2,000 HIV patients for a year asking them about the alternative therapies they are using and what the results are. This clinical trial has no control group of similar patients without HIV taking identical therapy treatments for comparison. The therapies themselves would not be controlled. Failure of this grant is guaranteed.

Dr. Thomas Kiresuk, psychologist at the Minneapolis Medical Research Foundation, plans to question centers for addiction treatment--for people addicted to cigarettes, alcohol and drugs--about which alternative therapies are most useful. Of course every institution is convinced of the usefulness of its own therapies or it wouldn't be in the business. In no way is this a scientific test.

And the list goes on.

The National Institute of Health has no monopoly on support of alternative medicine by a Federal Agency. The Department of Defense through the Uniformed Services University is funding a grant to Joan Turner, University of Alabama, Birmingham for $355,225 for a one-year study of "The Effect of Therapeutic Touch (TT) on Pain and Infection in Burn

Patients" [5]. A TT nurse will wave her hands over the burned patients without touching. A control group of patients will receive placebo therapy by mimic TT, by a person pretending to give the real TT therapy. The pain and healing of the two groups will be compared.

The TT theory is based on the idea that humans and the environment are filled with a 'life energy field.' The human energy field flows in balanced patterns in health but is depleted in illness. The therapist furnishes the human support system until the patients immunological system recovers. Present technology is not sensitive enough to measure the human energy field, they claim. But TT is sensitive to this energy.

Instruments to detect pressure, as well as light, sound, molecules we smell and molecules we taste have been developed that are far more sensitive than the human touch, sight, hearing, smell and taste. Yet, these instruments have never detected a 'medium' with the attributes of the life energy field that permeates all space. Just as the Traditional Chinese Medicine finds no support from science for the existence of Qi vital energy, there is no scientific evidence for the human energy field flow of TT.

Homeopathy reached its peak in popularity in the United States about 1900. It then dwindled away into near obscurity until revived by the 'New Age' in the 60s and 70s. Homeopathy is the alternative therapy that treats the patient by giving minute doses of a remedy that produces symptoms of the disease in a healthy person. It claims that the lower the concentration of the dose the greater the effect. Any chemist knows that is nonsense. The homeopathy logic can be carried to the extreme that no molecule of the curative agent remains in the dose. If there is no molecule of the agent there can be no cure.

I hope the examples of scientific and alternative medicine have persuaded the reader that the health of the world has, and in the future will be determined by scientific medicine. Alternative medicine will play a negative roll in human health until it is willing to subject its therapies to the rigorous scientific tests required of scientific medicine. Those therapies that survive will be supported enthusiastically by the scientific community.

In the columns that follow I discuss, primarily, topics that have been in the news. Too often, by default, the activists appear to be carrying the day. Scientists seem reluctant or too busy to explain their successes to the public. Publication in the scientific journals is important but that is not enough--the public expects more. I hope that the arguments given in the columns are convincing, and that they will assist the reader in understanding the issues and in making decisions that will enhance their health and the health of the world.

1. Fluoridation

Score a victory for scientific medicine. After 50 years of acrimonious opposition, California's Governor, Pete Wilson, signed into law a bill requiring the fluoridation of drinking water in cities with populations over 100,000. The requirement is effective January 1, 1996, except for those cities needing additional time to raise the costs.

The Fluoridation Bill, sponsored by Assembly member Jackie Speier, was approved by the Assembly Committee on Environmental Safety and Toxic Materials by a vote of 7 to 4.

The bill received bitter opposition despite support for fluoridation by the American Dental Assn., the American Cancer Institute, the American Medical Assn., the Centers for Disease Control, the Public Health Service, and the World Health Organization.

Opponents have claimed fluoridation is a Communist plot, violation of civil liberties and a promoter of cancer, birth defects, heart disease, AIDS and other diseases. However, numerous studies over several decades, including the U.S. Public Health Service in 1991, have found no credible evidence of any health risk.

I recall an election for fluoridation by our local water district in Danville, CA in the late 1950s. The issue lost when opponents from all over the state came to Danville to ensure its defeat.

As early as 1892, studies in England suggested that the lack of fluorine in the diet was responsible for tooth decay. The first to fluoridate its water was Grand Rapids, Mich., in 1945. In the study of 30,000 school children during the following 11 years, cavities were down by 60 percent.

Scientific tests since have shown that the easiest way to reduce a fluorine deficiency is by adding it to the drinking water to a level of about one part per million. At this level there is no risk to the body.

Three national surveys of children's' teeth in the 1980s confirmed the large reduction in cavities. The study in 1986-87 showed American children had 36 percent fewer cavities than in the early 1980s.

The ADA reports that because of fluoridation, 50 percent of American children entering the first grade have no cavities. It appears that children who drink fluoridated water from birth are most helped. And cavities in adults are reduced by 15 to 35 percent.

Fluoride has been successfully added to toothpaste, mouth rinse, gum, drops and tablets. It has been used to rebuild damage to enamel in teeth in the early stages of decay.

Seventy percent of U.S. cities with populations greater than 100,000 drink fluoridated water. However, only 17 percent of Californians fluoridate their water compared to 62 percent nationally.

In California--San Francisco, Long Beach, Oakland and Fresno have fluoridation. But Los Angeles, San Diego and San Jose do not.

Fluoridation is not expensive. Costs are estimated to be about 50 cents per person per year with a lifetime cost of about $40, a real bargain compared to the $50 charge of the dentist to fill one tooth.

The Goleta and Santa Barbara Water Districts both obtain water from the Cachuma reservoir. To assure water quality, Municipal Water Districts add certain chemicals to the water. When aluminum sulfate is added, a gelatinous aluminum hydroxide precipitates and removes fine sediment and bacteria from the water. Chlorine is added to disinfect as it kills many types of disease germs.

No fluoride is added to the water by the Goleta and Santa Barbara Water Districts. The natural fluoride concentration to the consumer is about 0.4 parts per million. So an additional 0.6 parts per million would be needed to reach the optimum concentration.

The fluoridation of municipal water systems has been one of the most successful public health programs ever carried out in the United States.

Available records indicate that neither the Goleta nor Santa Barbara Water Districts have ever held elections to fluoridate their water. Elections will not be necessary when the Speier Fluoridation Law takes effect.

2. AGING

I'll tell thee everything I can:
There's little to relate.
I saw an aged, aged man,
A-sitting on a gate
Lewis Carroll, "Through the Looking-Glass"

The early American statesman, inventor and scientist, Benjamin Franklin, will be forever remembered for his erudite comment, "In this world nothing can be said certain, but death and taxes."

Now even they are under attack, taxes by Congress and death by biogerontologists.

While Methuselah, an ancestor of Noah is said to have lived 969 years, the oldest confirmed person living today, perhaps ever, is Jeanne Calment, a French woman, 120 years old.

Life expectancy in the United States at birth increased from 49 years in 1900 to 76 in 1990. Seniors, 65 and older, constitute 12.5 percent of the U. S. population today and are projected to form 25 percent by 2050.

The increases in life expectancy during the early and middle 1900s came from improvements in sanitation, better maternal care and the prevention and curing of infectious and parasitic diseases with vaccinations and anti-biotics.

In the late 1900s, advances occurred in the early discovery and treatment of cancer, successful cardiovascular interventions and low risk life-styles. In the next century, life expectancy will continue to increase.

A convincing experiment has been carried out on fruit flies by Michael Rose, an evolutionary biologist at UC Irvine. With selective breeding he developed a stock of long lived robust flies having lifespans twice as long as normal flies. He suggested that the superflies produced an antioxidant enzyme, superoxide dismutase, that helped neutralize the free radical called superoxide.

Selective breeding was also used by University of Colorado biologist Thomas Johnson, who doubled the three week average lifetime of roundworms. He also observed that the average lifetime was increased by 70 percent with the mutation of only one of the worm's 10,000 genes, age-l.

Roy Walford at UCLA and others have demonstrated with mice that low calorie diets retard aging and increase length of life. This is found true also for most other animal species tested.

The cause of aging is still an open question. Is it by instructions from genes with programs that eventually command cell death? Or by damage from the highly reactive oxygen free radicals, fragments of molecules that remain after actions of the cell-immune defense and energy generating systems? Or by one of several other possible causes?

Aging is not a disease, but older humans and animals are more susceptible to disease. In the new science of biogerontology, scientists are studying the correlation of aging with the destruction and repair of cells and molecules in the body.

Although not yet demonstrated conclusively with humans, there is optimism that aging is governed by genes so may be slowed or stopped and even for a time, reversed. The growth hormone has been tested at V. A. hospitals with promising results.

Some scientists suggest that the amount of a hormone controlling genes goes down with age. That the added growth hormone rejuvenates anti-aging genes that cause the person to feel younger and stronger again.

The goal of aging research is to delay the aging process as much as possible and extend the time period of a happy, healthy, vigorous, productive life.

And yet increasing longevity poses possible serious social problems. Among the obvious are public support for retirement and medical and hospital care. As the population gets older, the burden on the rest will become unbearable unless corrections to public policy are made that are fair to all.

Some population experts think that the population could eventually approach a distribution with about the same number of people in each age group. That is greatly different from the present distribution concentrated at the lower ages.

The retirement age will have to be higher and the mix of jobs different in response to such an increase in life span.

3. Breast Implants

Help may be at hand if the courts follow the 1993 Daubert Supreme Court ruling. Judges should prevent plaintiffs from presenting scientific evidence to a jury unless it is "not only relevant but reliable." That is, anecdotal opinion is not sufficient. The scientific evidence must be 'falsifiable'.

Awards in breast implant settlements [6] may be "as little as 5 percent of the sums (women) were promised in the Proposed $4.2 billion settlement of the silicone breast implant lawsuits."

This is because many more claims were filed by the March 15 deadline than anticipated. While we all sympathize with the claimants who are suffering, no long term scientific study shows that breast implants cause a serious disease.

Silicone polymers are long chain molecules constructed from carbon, hydrogen, oxygen and silicon atoms. They are used in large quantities for oils, greases and rubbery solids.

Since 1950, silicone has been used for implants in small amounts for artificial joints, pumps, shunts and drains. Silicone is relatively non-reactive, non-allergic and usually tolerated by the body, so is used in pacemakers and in penile, testicular and breast implants.

Breast implants have been available for more than 30 years. In the United States it is estimated that over 300,000 women have more than 530,000 breast implants. The implant usually consists of silicone gel or salt water in an envelope of silicone.

Occasionally problems have occurred. These include rupture, leakage and distortion that usually can be corrected. However, others such as scleroderma, hardening or overgrowth of connective tissue, could leave long term damage.

Several advisory panels convened by the U. S. Food and Drug Administration concluded that, although more research was required, silicone breast implants should continue. In 1992, David Kessler, Commissioner of the FDA, overrode the advice and requested a voluntary moratorium on silicone implants.

Plaintiffs have been successful in suits against the implant makers and suppliers of materials. In 1994 a group of companies--including Dow Corning

Corp., Baxter International Inc., Bristol-Myers Squibb Co. and 3M Co.--agreed to a $4.2 billion settlement. The agreement now appears premature as the scientific studies of long term effects of breast implants are just coming in.

The latest is a study, "Silicone Implants and Connective Tissue Disease," of the Medical Devices Agency of the British Government, issued in December, 1994. The agency evaluated published data from the end of 1991 to July 1994.

The British Group concluded that "there was no evidence of any association between breast implants and connective tissue disease." They also found no reliable evidence for immunological problems related to breast implants.

S. E. Gabriel and L. T. Kurland, research physicians at the Mayo Clinic, Rochester, Minn., compared the medical records of 749 women with breast implants to 1,498 control women without. They concluded in their 1994 report, "We found no association between breast implants and the connective-tissue diseases and other disorders that were studied."

Other recent large studies with similar results were carried out by Brigham and Women's Hospital, Harvard Medical School, Boston, 1994; Johns Hopkins Medical Institutions, Baltimore, 1992; University of Maryland School of Medicine, Baltimore, 1993; and the M. D. Anderson Cancer Center, Houston, 1993. The latter found no correlation of autoimmune disease with breast implants.

Although there is no scientific evidence for association of breast cancer implants and disease why not play safe and ban silicon implants anyhow? It is because many patients are satisfied with their implants and others desire implants in the future. Other kinds of implants are also in jeopardy.

Litigation and fear of litigation has and is driving implant makers and material manufacturers from the market. The public may soon lose the advantages of the health implant devices. Among possible losses are angioplasty, cardiac catheters, cardiac defibrillators, blood filters and mechanical valves.

4. Repressed Memories

Some therapists claim they help patients recall repressed memories of childhood sexual abuse. Juries have handed down verdicts with the only evidence the testimony of the accuser. Reputations of family members and friends have been ruined. Innocent people were sent to prison. Are the repressed memories recovered or are they implanted?

Forensic DNA has recently been decisive in proving the innocence of men wrongly imprisoned. In each case the convictions had been obtained primarily on repressed memory testimony.

George Franklin Sr. was set free July 3, 1996 when San Mateo County prosecutors declared they had insufficient evidence to retry him for the 1969 murder of 8-year old Susan Nason.

Franklin was convicted in 1990 after his daughter Eileen Franklin-Lipsker, then 29, testified that while watching her own daughter, she remembered seeing her father molest and murder Susan.

Another daughter, Janice Franklin said she and her sister had been hypnotized to enhance their memories before testifying. This while knowing that testimony influenced by hypnotic suggestion is not admissible in California courts.

Franklin-Lipsker also told investigators in 1990 that she remembered her father committing two more murders. She remembered her godfather, Stan Smith, raping Veronica Cascio and her father murdering her. DNA semen tests proved neither Franklin nor Smith could have raped Cascio. Details of the other murder were too vague to investigate.

In a second case, Kevin Green was released from prison on June 20 after serving 17 years time for the wrongful conviction for beating his wife Dianna D'Aiello and killing their unborn 9-month old baby. At the trial she testified that she remembered Green attacking her.

However, the suspected serial killer, Gerald Parker, recently admitted the bludgeoning. DNA evidence has connected Parker to this and five other killings.

Child sexual abuse is a serious problem. Few crimes are as contemptible and crippling as incest and violence against children.

Memories of these crimes can debilitate children for years. There is no doubt that it occurs too often and public outrage is justified.

During the 1980's an epidemic occurred among women, mostly from 25 to 45 years old, who had "repressed memories" of childhood sexual abuse. The women had vivid memories, after clinical therapy sometimes for many years, of sexual atrocities by fathers, grandfathers, uncles or other relatives or close friends.

Several recent books including "Making Monsters: False Memory, Psychotherapy and Sexual Hysteria" by Richard Ofshe and Ethan Watters [7]; "The Myth of Repressed Memory" by Elizabeth Loftus and Katherine Ketcham [8]; and "Victims of Memory" by Mark Pendergrast [9] are critical of the repressed memory therapy. Martin Gardner's columns [10] in the Skeptical Inquirer have vigorously attacked this new "cottage industry."

Typically a woman will seek therapy for anorexia, anxiety, bulimia (abnormal craving for food), depression or headaches. Some therapists, not all, would soon, sometimes even during the first appointment, suggest that the symptoms were caused by repressed memories of abuse during childhood.

Shocked, the patient vehemently denies the diagnosis. But the therapist insists that she is "in denial" and suppressing painful childhood memories. Hypnotism, regression therapy or drugs such as sodium amytal are used. After many sessions of suggestion and persuasion by the therapist, the details of the acts seem real and the client remembers her abusers.

Dr. Loftus, a distinguished psychologist at the University of Washington, who has spent much of her career on studies of memory, says "The infancy memories are almost certainly false memories given the scientific literature on childhood amnesia."

The cited books and papers conclude that many therapists, instead of "recovering" memories, are "planting" them. Ofshe is a Professor of Social Psychology at UC Berkeley. In "Making Monsters," he and Watters explain, "Free from burden of proof, therapists have created an Alice in Wonderland World in which opinion, metaphor and ideological preference substitute for objective evidence."

There appears to be little scientific evidence to support "recovered memory.

5. Does HIV cause Aids?

Preliminary information was reported at the July, 1996 international AIDS conference at Vancouver, British Columbia that certain combinations of drugs may drive down the HIV infection in the blood of AIDS patients. Although limited optimism was expressed by some medical scientists, they cautioned we are still far away from a cure.

The number of cases of Acquired Immunodeficiency Syndrome (AIDS), reported to the World Health Organization by June, 1995, reached 1.2 million. Because of underreporting, the actual number is more like 4.5 million, and 20 million people are infected with human immunodeficiency virus (HIV).

Philip Rosenberg, in “Science” [11], estimates that about 1 percent of white, 3 percent of black and 1.5 percent of Latino males in the nation in 1993, 30 to 40 years old, were infected with HIV.

A patient with AIDS is unable to produce sufficient antibodies to fight life-threatening diseases like pneumonia. HIV is a retrovirus, a virus that can insert its own genetic material into that of its host.

The conventional wisdom, since the middle 1980s, is that HIV infection causes AIDS. Carriers of HIV are diagnosed HIV positive if they carry HIV antibodies. Most scientists involved in AIDS research hold this view.

However, there are a few dissenters, including Peter Duesberg, retrovirologist at UC Berkeley and his followers, who argue that AIDS is not due to HIV, but rather it is caused by illicit drug use and AZT, the much used anti-HIV medication.

Duesberg found enthusiastic support for his attack on the AIDS establishment from publications like the "London Sunday Times" which carried out a spirited exchange with John Maddox, editor of "Nature."

Because of the controversy, the journal "Science" carried out a three-month investigation of Duesberg's claims. The results were described by Jon Cohen [12]. He reported that although Duesberg "raises provocative questions, few researchers find his basic contention, that HIV is not the cause of AIDS, persuasive."

The investigation's main conclusions are that: copious evidence exists that HIV is the cause of disease and death in hemophiliacs (patients with hereditary blood-clotting problems); the AIDS epidemic in Thailand confirms the action of HIV; AZT and illicit drugs do not cause AIDS; and HIV satisfies the Koch-postulates for disease causation.

The Koch-postulates require the isolation of the microbe from the diseased organism (patient), be given to a healthy host (animal) where it causes the same disease and then the microbe be isolated from the diseased host.

The Centers for Disease Control and Prevention (CDC) listed 6859 AIDS cases caused by transfusions before the start of blood screening against HIV. Since then, only 29 cases have come from blood screened against HIV, a dramatic decrease. A study of the total membership of the British National Haemophilia Register from 1977-1991 found that the death rate among patients infected with HIV was about 10 times that of the uninfected.

Duesberg has argued that the contaminants of factor VIII, a clotting factor given to hemophiliacs, causes the immunodeficiency observed in AIDS. Those infected with HIV have received the most factor VIII. But studies by James Goedert from the National Cancer Institute found no association between dose levels of factor VIII and the likelihood of getting AIDS.

An indicator of the seriousness of infection in AIDS patients has been the concentration in the blood of T lymphocytes carrying the CD4 antigen. T lymphocytes are white blood cells that develop in the thymus and give a cell-mediated immune response to activation by a foreign molecule. The worse the infection, the lower the CD4+ count.

Two important articles [13], one by David Ho and co-authors and the other by Xiping Wei and co-authors, found that the T cells in infected blood were created only a few days prior to the measurements. The time interval was so short that only a small fraction of the cells were infected. The body's reaction to infection by HIV is very active, giving a high turnover of the lymphocytes and virus.

The debate continues with Duesberg and followers unwilling to acknowledge error. A new book by Duesberg with co-author Bryan Ellison "Inventing the AIDS Virus" came out in February, 1996. Peter Drotman from CDC says the book is not helpful because people acting on its messages could harm their health.

It is unfortunate that the disbelief in the role of HIV spread to the AIDS community.

Recent, 1997, lab and clinical research have settled the controversy. HIV definitely causes AIDS.

6. Medical Research with Animals

Medical scientists are doing their best to preserve our health and cure our diseases. Medical research is difficult, lengthy and expensive. It is a risky endeavor, with no guarantee of success. The animal activists would do themselves and the public a great favor to put their energies into supporting the best possible medical techniques--joining the scientists, doctors and volunteers like Jeff Getty at the frontiers of the latest research.

The animal activists have surfaced again with the recent transplant of bone marrow cells from a baboon to a human AIDS patient, Jeff Getty.

They have bombarded the newspapers with commentaries and letters to the editor denouncing the latest medical efforts to find a cure for AIDS. It is not just animal implants in humans they oppose but any use of animals in medical research to prevent or cure disease in humans.

These activists impede medical research locally as well as nationally. A year ago members of the Santa Barbara movement, Animal Emancipation Inc., were arrested for attempting to stop the delivery of rabbits to the UC Santa Barbara campus.

Animal activists claim medical researchers are torturers, imprisoners and murderers. Yet many of the surgeries are similar to those on medical patients at hospitals. The cages in research labs are like those in a veterinary hospital. And pounds and animal shelters and cars on streets kill hundreds of times more stray cats and dogs than do medical research labs.

Animal use in teaching and research is regulated by the National Institutes of Health and U.S. Department of Agriculture. Protocols for care and treatment of animals in university and private labs require considerate and responsible conduct. Surprise inspections and penalties create strong incentives for compliance.

Joseph E. Murray, 1990 Nobel Laureate in Medicine, reminds us [14] that polio was eliminated 50 years ago by a vaccine developed in monkeys and that insulin for diabetics was developed with research on dogs. The many antibiotics for pneumonia and other infectious diseases and chemotherapy for cancer were produced with the help of animal experiments.

Today, the development of drugs for cancer, AIDS, Parkinson's disease, schizophrenia and many other diseases is dependent on medical experiments with mice and other rodents. Genetic medical research, even behavioral problems such as alcoholism or obesity, often is first carried out with rodents.

Xenotransplantation--medical transplants between species such as animals to humans--is a new medical field with the potential of revolutionary changes in patient treatment. The activists complain that even if bone marrow transplants from baboons work, there are not enough animals to cure all the patients with AIDS. This is true, but the goal of the early experiments is to understand and demonstrate how transplants work. So in the future, the necessary cells can be harvested without sacrificing the donor. Eventually medical research should be able to grow the cells from cultures.

Thalidomide is a sedative that caused malformation of limbs of children born to mothers using it during pregnancy. Animal activists have perpetuated the myth that it was a well-tested drug. But the tragedy could have been avoided, if the drug had been tested on primates before use.

The human body is a very complicated biological system about which the medical research community has much still to learn. For example, the human genome, a complete set of chromosomes with the genes they contain, while under intense study, is still incomplete.

Medical research scientists use many approaches to solve medical problems. They employ animals in medical research only when there is promise of success. Replacing animal experiments with cell cultures, and animal biological systems with sophisticated computer programs is continually pressed by the activists.

These directions have been obvious to medical scientists for years as the laboratory techniques and computer capabilities have rapidly improved. Many medical researchers are already deeply involved in these types of research. However, cultures of cells are very limited biological systems. And sophisticated computer programs require, as prior input, a sophisticated knowledge of biological systems--data usually not available.

In most cases today, animal experiments are the obvious choice.

7. The Mad Cow Disease

The jury is still out on the cause of the Mad Cow Disease and a number of similar diseases in animals and humans. Are they caused by the prion protein or a phantom virus? Current evidence seems to favor the prion. The decision awaits the crucial experiment--making an infectious prion in the lab, or finding the elusive virus.

On March 21 the British Health Secretary, Stephen Dorrell, announced to the House of Commons that there may be a link between the Mad Cow Disease and the human Creutzfeldt-Jacob Disease (CJD).

The British Scientific Advisory Committee on Bovine Spongiform Encephalopathy (BSE), more commonly known as the Mad Cow Disease, found that 10 cases of CJD in the last 7 months, while not providing direct evidence of an association, were "cause for great concern."

Cows afflicted by the neurological BSE stumble, stagger around and eventually die. Spongelike holes are found in their brain tissue, much like in brains of humans suffering from CJD.

The BSE was likely transmitted to cows in the early 1980s by feed containing the brains and other internal organs of sheep infected with "scrapie," a neurodegenerative disease in sheep. That practice was stopped in 1987.

Due to alarm over BSE last fall, the sales of beef in Great Britain plunged 20 percent. And beef was banned in the British public schools.

Britains wondered if BSE could be transmitted to humans through beef in the diet and cause CJD. They asked whether new cases will show up in cows and humans, even though the feeding of scrapie to cows was stopped in 1987.

Their fears were partially removed in December, 1995 by a report [15] of John Collinge and his coworkers at St. Mary's Hospital in London. Treatments of mice genetically engineered to react to the agent that causes CJD in humans, remained healthy when injected with the agent causing BSE in cows.

While encouraging, this was not enough evidence to guarantee BSE could not be transmitted to humans.

So Mr. Dorrell's announcement set off near panic in Britain and other countries buying British beef. First France, Belgium and five German states banned imports of beef. Then the European Union ordered Britain to stop exporting cattle and beef products. Britain offered to destroy 15,000 cows a week for six years if the European Union would pay most of the cost. Details are still under negotiation.

The British Government has been criticized by European scientists for not communicating information sooner about the 10 cases. And the European Union has set up a high-level scientific commission to recommend research into possible links between BSE and CJD.

The 10 new cases of CJD are unusual. The average age of CJD victims is 62 years. The new cases average only 29 years. The usual CJD patients are diagnosed first with dementia and have a characteristic pattern of electrical activity in the brain. Not only was this absent for the 10 new cases but they were first diagnosed with psychiatric disorders such as anxiety and depression followed by problems in movement.

It appears that the 15 year old research of Stanley Prusiner, Neurologist and Biochemist, and his colleagues at UC San Francisco should have been taken more seriously. They pointed out the similarity of scrapie in sheep with "kuru" in humans in New Guinea highlanders and CJD around the world. They also gave convincing evidence that the scrapie agent was not a virus, viroid or plasmid but a small protein that he labeled a "prion," short for "protein for infection."

It could possibly be a small nucleic acid containing DNA and RNA surrounded by a tightly packed nearly impenetrable protein coat. Or perhaps a protein without nucleic acid, an infectious protein. Both were considered heretical. This should have been a warning that it might be possible for scrapie to cross species lines from sheep to cows to humans.

Scrapie and the Mad Cow Disease have not been a problem in America, although this isn't obvious from the movements of the stock market. The day after BSE was discussed on the Oprah Winfrey talk show, cattle prices fell precipitously on the Chicago Mercantile Exchange. It was anticipated viewers would stop buying beef.

How bad is the risk? CJD occurs in about one out of a million people each year. The 10 new cases in Britain give an added 2 per 10 million. The risk of death each year in an automobile accident is about one in 10,000, a risk often taken without thought.

Unless the 10 cases are just the tip of the iceberg, Britains can continue to dine out on prime rib without worry--and hopefully survive the drive home.

8. Electromagnetic Fields and Cancer

In April, 1995 the American Physical Society Council stated that purported health effects of power line electromagnetic fields have not been scientifically substantiated and the cost of mitigation and litigation "is incommensurate with the risk, if any."

The Santa Barbara County Board of Supervisors voted unanimously on April 26, 1995 for a sensible approach to the alleged hazards of Extremely Low Frequency Electromagnetic Fields (ELF-EMF).

Developers will not have to submit environmental reviews and special regulations will not be required for housing, schools and day-care facilities close to high voltage power lines. Development standards will be adopted in a few weeks.

Earlier, harsher regulations were considered in response to parent concerns when seven Montecito children, five in the Montecito Union Elementary School, developed leukemia and lymphoma between 1980 and 1989. An electrical substation and a high voltage power line sit near the school.

Cancer in children is a tragedy. But it cannot be prevented by assigning the wrong cause. The evidence is that power lines, transformers, ovens, heaters, toasters, curling irons, TV and computers are not responsible for cancer in children and leukemia and brain cancer in adults.

While a few scattered studies in the United States and abroad have suggested a connection between cancer and ELF-EMF with low statistical reliability, most have not. There is no smoking gun--no obvious connection like the badly destroyed hands of the early x-ray technicians.

The most extensive study to date is the June 1992 report [16], "Health Effects of Low-Frequency Electric and Magnetic Fields", prepared by the Oak

Ridge Associated Universities (ORAU) Panel. The panel consisted of 11 members selected from the fields of bioengineering, biology, epidemiology, medicine, physics and statistics.

The panel reviewed about 1,000 journal articles published in the last 15 years. They worked individually and in groups and brought in additional expert consultants as required.

The panel concluded, "there is no convincing evidence in the published literature to support the contention that exposures to ELF-EMF generated by sources such as household appliances, video display terminals, and local power lines are demonstrable health hazards."

Associations between electric and magnetic fields and child and adult leukemia and cancers are inconsistent and inconclusive. No biological mechanism has been suggested to explain a causality. No conclusive evidence exists that the fields initiate or promote cancer or influence tumor progression.

There is no convincing evidence that electric or magnetic fields cause birth defects or reproductive problems, and any neurobehavioral effects are temporary without health consequences.

In 1994, William R. Bennett Jr, Professor of Engineering and Applied Science and of Physics at Yale University and a member of the ORAU Panel, published [17] his book, "Health and Low Frequency Electromagnetic Fields" and an article, "Cancer and Power Lines," in "Physics Today."

He brought the ELF-EMF studies up to date, affirmed the conclusions of the panel and directed attention to the electric and magnetic fields that occur naturally inside and outside the body.

Natural and man-made electric and magnetic fields are always in and around our bodies. The static background electric field outside is about 120 volts/meter and the magnetic field about 0.5 Gauss. Beneath high voltage transmission lines at head level, alternating electric fields are less than about 10,000 volts/meter.

Since the body is a good conductor, the field inside the body is reduced by a factor of 1 to 100 million. The resulting field is negligible compared to the normal background electric fields in the body from thermal fluctuations. Although the external magnetic field is not reduced in the body, the induced electric field at low frequencies is also small compared to the background thermal electric field.

D. Allan Bromley, Science Advisor for President Bush, estimated that the ELF-EMF cancer panic has cost the US public a non-trivial $23 billion since 1989. That would go a long way toward finding the real source and prevention of cancer.

9. Nuclear Medicine

Many Americans owe their lives to nuclear medicine. For that reason it is strange that so many consider Nuclear and Radioactivity as the N and R words of today. This column may remind us of some of the benefits of nuclear medicine.

One of the best kept secrets in medicine is the high use of radioactive isotopes to detect health problems and treat them. Radioactive isotopes have made significant contributions to the health of most Americans.

The pioneering work in nuclear medicine was carried out by John Lawrence, brother of Nobel Laureate E. O. Lawrence, at the Crocker Lab on the UC Berkeley campus. The early cyclotrons produced the isotopes for use in the medical studies.

Glenn T. Seaborg, University Professor of Chemistry at the University of California and Associate Director-at-Large at the Lawrence Berkeley Laboratory, has long been a champion of nuclear medicine. He was also co-winner of the 1951 Nobel Prize in chemistry.

He points out the news [18]—startling to many--that in the United States, ten million patients per year, one-third of all those hospitalized, are treated with nuclear medicine and more than 100 million nuclear medicine treatments are prescribed each year.

For diagnostic purposes, a chemistry compound containing the desired isotope is injected into the bloodstream and moves to the target organ. The isotopes decay into gamma rays that are detected by external gamma ray cameras.

From the many isotope decays, the location, shape and other properties of the tumor can be determined. Or after a cancer is located, it can be destroyed by external sources of radiation.

Some of the isotopes most used today, including cobalt-60, iodine-131 and technetium-99m (the name gives the element and the number the isotope), were discovered by Glenn Seaborg and his coworkers at Berkeley.

Technetium-99m is the most widely used isotope for diagnosis of medical problems. When a small amount is injected into the bloodstream it migrates to various organs of the body that are imaged by gamma-ray cameras. Each year over seven million scans and images are taken of bone, liver, kidney, spleen, lung, thyroid, and brain in search of tumors and other abnormalities. For example, the spleen can be checked for rupture and internal bleeding. For these diagnostics, the radiation dosages are low.

Technetium-99m has a 6 hour half-life, too short for shipping from nuclear reactors where it is produced. Instead its parent, molydenum-99, with a half-life of 66 hours, that decays into technetium-99m, is shipped in container 'cows' that are 'milked' every few hours to obtain the technetium-99m.

The half-life is the time for half of the isotope nuclei to decay. After two half-lives, one-fourth remains; and after three lifetimes, one-eighth. So after 10 half-lives only about one-thousandth of the original isotope remains.

Thus isotopes like Technetium-99m quickly disappear.

Cobalt-60 is widely used for destroying cancer with little damage to surrounding tissue. More than four million cobalt-60 treatments are carried out each year. For maximum damage to the cancer, larger dosages are required.

It is also possible to bombard cancer with beams of protons or alpha-particles. These charged particles have the advantage of higher energy deposits in cancerous tissue near the ends of their paths .

As iodine concentrates in the thyroid, iodine-131 can detect and treat thyroid problems. Former President George and first lady Barbara Bush were both treated successfully with iodine 131 for Graves' disease.

Nuclear medicine has relieved the pain and prolonged the lives of millions of people throughout the world. It has played a major role in the increase in life expectancy of Americans at birth from 49 years in 1900 to 76 in 1990. In the years ahead we may expect nuclear science to contribute to the welfare of all people, not only in nuclear medicine but in many other important aspects of their lives.

10. Traditional Chinese Medicine

Traditional Chinese Medicine (TCM) theory and practice is little changed from that of the ancients in the third millennium BC. Their theory of life energy is not supported by modern science. Some TCM treatments like proper diet are normal scientific medicine practice. Others like acupuncture need further scientific study. There is no scientific evidence for TCM paranormal claims.

Alternative medicine seems to be gaining popularity in America these days.

Examples are homeopathy, touch therapy, aroma therapy, naturopathy, acupuncture, moxibustion and herb medicines. The latter three are practiced in Traditional Chinese Medicine or TCM that originated in China over 4000 years ago.

TCM teaches [19] that "life energy" or "Chi" or "Qi" permeates all space and runs through the body along lines called "meridians." Acupuncture therapists stick needles into the skin at designated points to alter or control the energy flow. As many as 600 points on the body are used in healing.

Stimulation of the points by hand and by heat from "moxibustion" are also considered beneficial. The herb, moxus, is normally burned at the appropriate point on the skin and smashed to form a blister.

Other techniques for increasing the flow of Chi are special diets, exercises, massage and herbal medicines.

Through the practice of TCM, it is also believed possible to perform superhuman feats, carry out paranormal acts such as telepathy and telekinesis and protect oneself from injury.

Acupuncture-anesthesia was used for the operation on the inflamed appendix of "New York Times" reporter, James Reston in China in 1971.

The vital force or field is not included among the gravitational, electromagnetic, weak and strong forces of nature. The claimed properties of the Chi field violate the existing theories of physics, chemistry and biology.

Scientists have used extremely sensitive instruments to search for the energies and fields of nature including the human body. The energies and fields of physics, chemistry and biology have been detected and precisely measured, but no energy with the properties attributed to Chi has been detected. And the instruments used are far more sensitive than the human senses of touch, sight, smell, hearing and taste. There is no scientific evidence for the existence of Chi.

Then how do we account for the success of TCM in curing certain illnesses?

Some of the practices such as diet, exercise, massage and certain medicines find support from scientific medicine. Scientific theories have been proposed for their success and controlled tests have been carried out to demonstrate their effectiveness as cures.

There is little doubt that some people are less sensitive to pain than others. Also reaction to pain, like many other bodily responses, can be conditioned with practice. There still seems evidence that acupuncture-anesthesia does work. Among possible scientific explanations is the "gate theory." It suggests that the gentle insertion of the needle stimulates a nerve that blocks the passage of a stronger pain signal.

The 1988 Skeptical Inquirer team [19] in China was able to test some of the key claims of TCM. A Qigong master, Dr. Lu claimed that Qi emanated from his fingertips and that the patient responded to his movements. He demonstrated this to the team in a room with the patient on a couch. He also claimed his therapy worked even if the patient was in another room and said 15 seconds was long enough to transmit his Qi.

With the patient in another room a coin was flipped. If heads, Dr. Lu transmitted Qi for 15 seconds, if tails he did nothing. The patient was watched by observers who took notes of each 15 second interval. The patient agonized during the whole trial independent of therapy from Dr. Lu. When in the same room, it was clear that Dr. Lu's actions followed, not led, the patient's movements.

At the Medical Institute of Sichuan Province it was claimed that 16 children had paranormal abilities--they could identify objects hidden in closed envelopes or match boxes. When precautions were taken against cheating, none were successful.

CHAPTER IV

THE ENVIRONMENT

Over the last few decades, the environment has become an important issue with the American and world-wide public. Pollution of the earth's air, water and soil and protection of animals and endangered species is a major concern.

Hundreds of conservation and environmental groups, councils, leagues, institutes, centers, alliances and societies have been established to plan, protect, defend, save and fund wildlife, parks, the earth, forests, wetlands, endangered species, animals, birds and the wilderness. Each cause has its dedicated supporters and activists. Their opposition to the establishment, technology and science seems always present. The cacophony of the activists grows ever louder. Many examples could be cited.

A membership solicitation letter of the Union of Concerned Scientists (UCS) cries out, "America is addicted to oil. And we are paying dearly to feed our habit. We are paying with oil spills, polluted air, and global warming........ Research by the UCS has shown that the science and technology are available that could stop the destruction of our environment....."

The UCS is referring to renewable energy--solar, wind and biomass. These have never produced more than one percent of the nations energy and show little promise of improving in the future. Biomass would be as polluting as fossil fuels and solar collectors and windmills are sight and sound polluters. However, nuclear reactors are completely contained, send no pollutants into the atmosphere and the used reactor fuel rods can be stored safely in high radioactive repositories like Yucca Mt. in the Nevada Proving

Ground. The Union of Concerned Scientists has always been opposed to nuclear reactor power.

The California Air Resources Board in March, 1996 revoked the mandate that two percent of the cars purchased in California in 1998 must be electric battery propelled with increasing percentages through 2002. Joseph Caves, lobbyist for UCS said, "The ARB truly believes that they can trust the car companies to develop this technology and put it into the market. I don't think they've earned that kind of trust."

He is wrong. It's not a matter of trust. It's a matter of science and technology that has not yet been developed. It finally became obvious to the ARB that cars with batteries available in 1998 would not be capable of giving the distance of travel, acceleration and convenience at a price the public would pay. It is very likely this will still be true in 2003 and later. Once again it was demonstrated that science and technology cannot be mandated.

A membership drive letter from Jan Beyca, National Audubon Society senior scientist, reads, "The destructive processes man has set in motion resemble a cancer. .....We can project with some accuracy the eventual end of the natural world as we know it. That is, no trees. No wildlife. Climate changes so radical the tropics have migrated to the North Pole......Yes people will survive. But they will live in the future equivalent of caves, insulated from the environment."

Paul Ehrlich, Stanford University ecologist and a leading advocate for population limits, has done the world a favor with his arguments against overpopulation. But he has also ruined his scientific credibility with his many doomsday predictions that never came true. In 1968 he said, "The battle to feed humanity is already lost.....we will not be able to prevent large-scale famines in the next decade." Instead, since 1968 world grain production has increased 60 percent. And the rate of food production has gone up faster than the rate of population growth. In 1980 Julian Simon, economist at the University of Maryland, bet Ehrlich that over 10 years the price of certain scarce metals would fall as increased demand spurred production. Ehrlich maintained the scarcity would drive prices up. Simon won the bet.

David Brower, long-time environmentalist, 15-year-president and now board member of the Sierra Club, reports [1] he is distressed by President Bill Clinton's environmental record. Brower is critical of the "salvage logging" rider compromise arranged by Clinton to permit some logging in the northwest, earlier stopped by the spotted owl. However, many residents, perhaps most, are opposed to the rider for a different reason--they think it is too restrictive, it doesn't allow enough logging.

Brower is opposed to "the weakening, if not the gutting, of the Endangered Species Act through administrative changes in its rules and regulations." But many citizens think the act should be modified to consider habitats suitable for multiple species and give some protection to landowners. Brower is opposed to "opening wildlife refuges to hunting and fishing by presidential decree; .... weakening the Safe Drinking Water Act by allowing increased levels of lead and arsenic in drinking water supplies; .....and reversing the ban on the production and importation of polychlorinated biphenyls (PCBs)." Yet each of these changes has been recommended by panels of scientists who have found they will correct current inequities and will not further endanger the environment, animals or people.

Greenpeace is a militant environmentalist organization. In 1995 it challenged the Goliath France in weapons tests near Mururoa Atoll in the South Pacific. The protest was stopped only after French navy commandos seized their ship, the Rainbow Warrior II.

The activist group also opposed the Royal Dutch Shell plans to dump the obsolete Brent Spar oil platform into the North Sea. Royal Dutch Shell scientists presented convincing arguments that less damage to the environment would occur that way than by dragging it to dry land and dismantling it there. Under extreme environmentalist pressure Royal Dutch Shell caved in even after John Major, Britain's Prime Minister, defended the plans.

At issue were the 100 tons of tar and hundreds of pounds of toxic heavy metals on the platform. The Shell scientists knew that organisms on the ocean bottom had adapted to similar materials from hydrothermal vents that furnish their energy. By enabling these bacteria to feast on the tar and toxic heavy metals, pollution could be safely avoided.

To Greenpeace's credit, it later apologized saying it had incorrectly estimated the amount of tar and toxic waste. The Brent Spar oil platform was eventually toppled onto the ocean bottom.

Earth Day is a day of celebration for the environmentalists. Supporters carry banners, "WE ARE ALL BEING POISONED," and "THE PLANET IS DYING." It is a family day of fun and play. Small children carry balloons and play in the sand. Youth lie on the grass listening to the rock band's heavy beat against the earth's desecration. More than 100,000 people celebrated the 25th anniversary of Earth Day in 1995 on the National Mall near the Capitol in Washington, D. C.

How many celebrants really know what the earth is like today, or was 10 years ago? Most are too young to have lived during World Wars II and I. How well do they know the world history of the last few centuries or of the dark ages and before? Do the protesters have some knowledge of the earth ten thousand years ago at the beginning of human civilization, or one million years ago (mya) at the time of the origin of the human species?

Could those at the festivities compare the earth today to its appearance at the time of the dinosaurs 100 mya--or to the earth at the time of the origin of life, about 3.8 billion years ago (bya)? And how many could describe the earth at its origin about 4.5 bya?

FORMATION OF THE EARTH

There is good scientific evidence that the Big Bang Origin of the universe occurred about 15 bya. Galaxies began forming about 1 billion years later. The stars are born in dense nebula, evolve and die, often as supernova explosions. Chemical elements heavier than hydrogen and helium are manufactured in the process.

The sun is one of the newer stars. Radioactivity measurements in the crust of the earth and the surface of the moon tell us that the earth is about 4.5 billion years old. Astrophysicists believe the sun and planets were formed when a giant rotating nebula of gas, mostly hydrogen collapsed into a disk. About 99 percent of the gas was drawn by gravitational force to the sun. Most of the rest was captured by the outer planets. The earth has less than one-hundred-thousandth of the mass of the sun.

The temperature at the center of the sun is about 10 million degrees Kelvin (one degree Kelvin is 1.8 degrees Fahrenheit) hot enough to burn hydrogen to form helium by fusion nuclear reactions. The energy diffuses to the surface of the sun where it is radiated to warm the earth and other planets.

It is thought that the planets formed by accretion--building grains of silica and carbon compounds by collisions. The few centers that grew most rapidly became planets. Slower growing ones ended as asteroids and comets. The inner planets--Mercury, Venus, Earth and Mars--never became massive enough to attract the remaining nebula gas of hydrogen and helium by gravitation. The outer planets Jupiter, Saturn, Uranus and Neptune were farther away from the sun and thus cooler. They grew larger and attracted the gas so have compositions similar to the sun.

Considerable debris of small meteoroids, asteroids and comets remained in the disk. The planets continued to grow as the objects collided and stuck. During this stage considerable water from ice in the comets and organic matter was added to the earth--crucial for the later formation of life. It is thought that this very violent period of evolution came to an end about 4.4 billion years ago.

Evidence of continued bombardment at a lower rate was obtained from pictures of craters on the moon, and other planets and their moons, by telescopes on the earth, the Apollo manned space program and unmanned probes to the distant planets.

Measurements of earthquake generated seismic (sound) waves that pass through the earth determine the structure of its interior. An inner core of high density iron is located at the center of the earth. It is surrounded by an outer core of liquid iron. The outer core carries the currents that produce the earth's magnetic field.

Surrounding the outer core is the mantle, a major part of the earth, that extends from the crust down to a depth of about 3,000 kilometers (1.6 kilometers equals one mile). The temperature at the base of the mantle is about 4,500 Kelvin (one Kelvin equals 1.8 degrees Fahrenheit). We live on the top of the crust, the thin outer layer of the earth only 20 to 70 kilometers thick.

The crust has been changed in many ways by bombardment of objects in the solar system from above, and by volcanoes and earthquakes from below. You have probably visited Meteor Crater close to highway 40 in Arizona. That crater, about a mile in diameter, was created by a million ton meteoroid 50,000 years ago. Because Craters erode, many earlier larger ones are no longer visible, except sometimes to the experts.

Most of the geologic features on the surface of the earth, the mountains, valleys, volcanoes and continent were caused by Continental Drift.

As a fourth grade student in geography you probably looked at a map of the world on the wall and saw the similarity between the shapes of the coastlines of Africa and of North and South America. They seemed to fit together like a jig-saw puzzle. If before 1960 you may have thought, maybe they once were one continent but split apart and moved away from each other. "No," your teacher might have said, "It couldn't have happened. The matching shapes is just an accident."

When Alfred Wegener, a German meteorologist in the early 1900s made the same observation, he got a similar reaction from the geologists. They said it was impossible; they thought the crust and the mantle formed an unbreakable solid. However, new evidence was found for Continental Drift in the 1960s; the continental crust and top mantle, about 60 kilometers thick, seems to float like a ship on the top of a mantle-ocean. Hot lava rises from the interior of the mantle, at ridges in the ocean where the crust is weak, and drives the plates apart. The plates travel at speeds of a few centimeters a year (one inch is 2.5 cm). Thus the Atlantic Ocean was formed in about 100 million years.

Evidence for the continental drift is abundant. Lava near the ridges was younger than lava farther away. Little iron magnets on the ocean floor were aligned along the direction of the magnetic field that existed at the time the molten lava solidified. The magnetic field of the earth changed direction every few thousand years and the changes of directions of the magnets on the ocean floor were in good agreement. Comparisons of species of flora and fauna of Africa and South America concurred with predictions of the drift.

But if material rises from the mantle and spreads the ocean floor it must also return to the mantle somewhere else. That happens at ocean trenches where one continental plate runs over another. Material from the lower plate returns to the mantle again. At these trenches the crust is also pushed up to form mountain ranges. The Sierra Nevadas in California were formed when the North American plate slid over the Pacific plate. The Himalayas are at the boundary between the Indian and Eurasian plates.

These plate boundaries are locations of violent seismic activity. Earthquakes shake the earth as the earth realigns along faults. Volcanoes spew hot lava down mountain sides and vent toxic fumes into the atmosphere.

The toxic gas and small ash particles sometimes remain in the atmosphere for years and even change the climate. Volcanic activity formed the Hawaiian Islands. Other volcanoes in the ocean nearby are currently creating new islands.

Erosion of mountains and plains by wind, water and ice change the contours of the earth and wash soil into valleys, rivers and oceans. Glaciers slice out new valleys as they move down mountain slopes. Rivers carve gorges and canyons through solid rock. The face of the earth is continually changing.

THE WEATHER

The Earth's atmosphere is about 80 percent nitrogen, 20 percent oxygen and one percent argon with traces of water and carbon dioxide. Oxygen breathed by humans and other animals combines with food we eat to furnish the energy for life. The reaction products include water and carbon dioxide. The plants absorb carbon dioxide and release oxygen in the photosynthesis of light. The cycle is completed as the oxygen replaces that consumed by humans and other animals.

The Earth's atmosphere was formed from the outgassing of the interior of the earth. Large amounts of hydrogen and oxygen are stored in the oceans as water. Considerable carbon dioxide is stored in carbonates in the crust and the ocean that may be released on heating.

Airplane travelers know that planes fly at about 35,000 feet to get above the weather that is caused by the movements of air in the atmosphere and evaporation and condensation of water. When water evaporates, large quantities of energy are stored in the water vapor. This energy is released as the water vapor condenses, causing violent storms with thunder and lightning and occasional hurricanes. The rapid rotation of the earth produces storms in the temperate zones which sometimes develop into tornadoes.

Hurricanes frequently devastate the Bermuda Islands, the southern Atlantic and Gulf coasts of America and many other parts of the world. Superstition and legend had given the Bermuda Triangle the reputation as one of the most treacherous regions whether traveling by air or by sea. But satellite tracking of the hurricanes and early warnings in recent years--along with strengthened building codes in hurricane zones have reduced the toll in lives and damage. For this we can thank science and technology.

Tornadoes are the plague of the Midwestern and southern states. The early prairie settlers had storm cellars where they stored their food and escaped from the terrifying tornadoes. As the tornado funnels swept along the ground, entire communities and even small towns have been wiped out.

Most of us have experienced floods that swell the rivers, flow into cities and cover large areas of fertile farms. Lives are lost, property is damaged and crops are destroyed.

Droughts are serious in many parts of the country where farmers depend on rain for their animals and crops. In dry regions without rain for many months of the year, grass and forest fires endanger lives, animals, crops, forests and homes.

Typical destructive weather is described in a recent Los Angeles Times diary [2]. Floods in central and southern China killed 1,400 people in recent weeks. Areas in Chicago are recovering from the worst flooding on record. Ten people died and 10,000 others were evacuated from their homes in northern Quebec Province.

Forty-three buildings were damaged by a magnitude 6.6 temblor on the Indonesian island of Sulawesi. A five month old drought in Ecuador forced more than 3,000 people from their homeland in the Andes Mountains. Huge waves from a tropical storm south of Tahiti damaged 400 homes in French Polynesia. Shores south of Papeete and Moorea were washed by tides 17 feet higher than normal.

Mt. Etna in Sicily spilled streams of molten lava down the mountain creating a brilliant display of natural pyrotechnics. Typhoon Frankie killed 14 people in northern Vietnam. Tropical storm Gloria strengthened to near typhoon force near Luzon, Philippines, then drenched Taiwan and eastern China already severely flooded.

The weather is not benign. It is driven by tremendous forces. The best efforts of meteorology, other sciences and technology are needed to predict the weather, take mediating action and avoid its devastating consequences.

EVOLUTION OF PLANTS AND ANIMALS

Earth Day also seems an appropriate time to look around at the plants and animals, and ask how they have evolved in time. From fossils of plants and animals and the radioactivity in the surrounding material, paleontologists have been able to date significant steps in evolution.

The earliest life, about 3.8 bya, started with the prokaryotes, bacteria whose genetic material is not contained in the nucleus of the cell. Following at 2.5 bya were the eukaryotes, organisms whose genetic material is inside the cell. Most life is of this form. Divisions of the animal kingdom began during this period. By 600 mya many animal divisions had appeared and there were diverse forms of algae. From 500 to 450 mya many animal divisions expanded and the first jawless fish appeared. A mass extinction ended this period.

From about 440 to 350 mya the first bony fish emerged and dry land was populated by plants and animals, later by insects and amphibians. A second extinction occurred near 350 mya.

Next vast forests covered the land. Reptiles and insects rapidly multiplied. A third mass extinction, especially of marine life, ended the Permian period about 250 mya. Ninety percent of the animal species were annihilated. Andrew Knoll of Harvard University's Botanical Museum and three colleagues in the journal Science suggested [3] that the extinction may have been caused by the up-welling of large quantities of carbon dioxide dissolved in ocean water--carbonated water. Carbon dioxide gas was released that blanketed the land and by the "greenhouse effect" raised earth surface temperatures. The natural habitats were so altered that many species could not survive. The marine life could have suffered from acidosis, too much carbon dioxide dissolved in the blood. It is possible that some of the other extinctions could be explained the same way.

Up to this time the continents were together as one large land mass. They then started drifting apart. Early dinosaurs and animals appeared. A fourth mass extinction was recorded about 200 mya. During the following Jurassic period dinosaurs multiplied and diversified. The first birds appeared.

By 140 mya the continents had widely separated, flowering plants and mammals diversified and the dinosaur was king. At about 60 mya the dinosaur era abruptly ended. Evidence is good that a large asteroid struck the earth near the Yucatan Peninsula in Mexico. A huge cloud of dust containing sulfur dioxide and other compounds blanketed the earth for several years. This likely changed the climate and destroyed the dinosaur food. The dinosaur population never recovered.

The mammals, birds, flowering plants and pollinating insects continued to diversify. The continents drifted close to their current positions. About one mya the human species evolved. Repeated glaciations took place. And many large mammals became extinct.

According to Darwin's Theory of Evolution, people and other animals and plants evolved from the original single cell organisms present at about 3.8 bya. Organisms evolve by natural selection. Selection depends on the variability in organisms. The variations are provided by mutations of genes that occur by chance. Most mutations are detrimental so that organism dies. Occasionally the mutation gives the organism an advantage so it has a better chance to survive and to pass its genes, including the mutation, to future generations.

Life is a continual battle for existence. The animal or plant must have the right environment to survive. Plants must have the proper climate, water, certain elements for nutrition, carbon dioxide and sunlight. They must have natural resistance to plant diseases, pests and fungi. Animals need climate, water, proper food and oxygen. They too must have resistance against disease. As part of the food chain, to survive, each animal must be able to avoid its natural enemies. In any habitat the number of survivors fluctuates continually with populations depending on the above factors. The larger the habitat the easier the animal or plant finds conditions satisfactory for its survival.

Until a few hundred years ago only about 100 million people lived on the earth. The population had been nearly constant for several thousand years. But with better sanitation, scientific medicine and healthier living habits the population skyrocketed to the current six billion people.

Science and technology have been so successful that life expectancies have or will soon reach 80 years for people in the developed countries. The success of humans has put a new stress and burden on other animals and plants. That is the concern of the environmentalists and ecologists. It is a concern we all share and its correction must be given high priority.

The obvious solution is to limit the world's population. Family planning, birth control, contraceptives and abortion all contribute, and are being successfully applied in many parts of the world. However, there are large pressure groups opposed to some or all of these methods that make population limitation difficult. Among the groups are various religious denominations including the Catholic church and several conservative Protestant denominations.

Some of the underdeveloped nations in India, Africa, Asia and South and Central America with large agricultural populations historically have had large families. The many children were a help in farming the land. With new scientific methods of agriculture and technological improvements in farm machinery, large families are not necessarily an advantage. Nowadays, there seems to be an inverse relation between the economic status of the family and the number of children. The populations of many of the developed countries have dropped below the sustainable levels and their total populations would be dropping except for immigration.

The latest estimates seem to indicate that the birth rates are going down in the underdeveloped nations fast enough that the world should reach a steady population by about the year 2050. Of course, that steady population will be much higher, perhaps 10 billion or more, because the world must wait for the replacement generation to die.

Environmental Policy

We now return to the pressing problems of today. Those of the pollution of air, water and land, of endangered species and food safety that cannot wait. What has been done to solve these problems?

Congress, realizing the need for management of the environment, in 1969 passed the National Environmental Policy Act. It established the Environmental Protection Agency (EPA) with the responsibility for overseeing environmental problems and setting standards.

Philip Abelson, Deputy Editor of Science, reviewed [4] the difficulties caused by the new regulations. By 1990, the regulatory machinery had become so complex that EPA had to administer 11 major statutes with over 9000 regulations. In 1993, there were 125,000 federal employees at EPA and other agencies busy creating new regulations. The direct costs each year to meet the mandates was estimated to be $500 billion, with indirect costs an additional $500 billion.

Initially, EPA targeted large chemical companies--later cities and towns. Water supplies were to be monitored for more than 130 chemicals, some in the range of one part in a billion. More recently EPA has aimed enforcement at small businesses. One regulation requires businesses to keep inventories of uses of 328 chemicals that must be reported to local and federal authorities. Many of the regulations have little effect on pollution or health of employees and are very expensive for small businesses to comply.

Since the Superfund was established in 1980 to clean up the worst hazardous sites, only a few have been restored. The laws seem to encourage litigation rather than remediation. Typically it takes 10 years and $30 million to restore a site. We need to ask the question again and again. What benefits do we get for the costs?

Abelson is especially critical of EPA's assessments of risks at Superfund sites where hazards have been exaggerated by factors of 100 or more. Abelson says, "EPA has set maximum concentration level goals of zero for some major chemicals of doubtful carcinogenicity. By law, cleanup levels of Superfund sites must under certain circumstances meet standards set by EPA under the Safe Drinking Water Act. Thus EPA has an invitation to require expenditures of trillions of dollars at sites around which few if any excess deaths have been seen." The nation cannot stand such costs.

Economics and endangered species have been on a collision course. In southern California development has been held up on 100s of square miles of land for more than six years by the kangaroo rat in Riverside County and the gnat catcher songbird in San Diego County. Millions of acres of forest in Washington and Oregon have been restricted for logging to preserve the habitat of the spotted owl. As the population increases there is more competition for land. Farmers, miners, loggers, cattle raisers, home owners, recreationalists, and environmentalists who want no change, battle for the use.

A major turning point in the developer-endangered-species-stalemate nationwide may recently have been reached. In southern California, a plan for satisfying both development and endangered species was agreed to by Irvine Co. Chairman Donald Bren, U.S. Interior Secretary Bruce Babbitt and California Resources Agency Secretary Douglas Wheeler. In July 1996 they signed a pact for a 37,000 acre wildlife preserve in Orange County. The plan protects the gnat catcher and nearly 40 other species by preserving the native habitat of coastal sage scrub from Costa Mesa to San Juan Capistrano. Development will be prohibited for at least 75 years. Landowners who participate are freed from the strict endangered species laws on land outside the protected area--but of course must satisfy the other development regulations.

Success resolving environmental conflicts is achieved by persuasion and cooperation, not by in-your-face tactics. Ralph Regula, Republican Congressman for 24 years from Stark County, Ohio has an uncanny ability to bring industrialists and environmentalists together on environmental issues. In the 1960s he was instrumental in helping the Timken Steel Company and the Audubon Society found the Wilderness Center. It is a 573 acre nature preserve in Wilmot, Ohio. Activists and steel executives alike go there to hike in the woods and meadows. Teachers with classes of all ages from the surrounding communities visit to learn about nature.

Congressman Regula in the 1970s worked with the Democratic Governor John Dilligan to create the Ohio Environmental Protection Agency. He and Representative John Seiberling, Democrat from Akron formed the Cuyahoga Valley National Recreation Area, a park in northeast Ohio, used by more than a million outdoor visitors a year.

After almost two years of bitter wrangling over EPA and the environment between the Republican-controlled House and Senate and the Democratic President--suddenly just before the party nominating conventions in the summer of 1996--bills were passed and laws were inked.

President Clinton signed the Food Quality Protection Act, the long overdue revision of the Food and Drug Law. It repealed the 1958 Delaney clause that banned any processed food that contained a detectable amount of cancer-causing pesticides. Detection had become so sensitive that foods with concentrations of less than one part in a billion, clearly not a danger, were being banned. Furthermore, much higher concentrations of natural plant pesticides of many parts per million were not subject to control.

The newly allowed concentrations of pesticides is set at a risk of one chance in a million of causing cancer in a lifetime of exposure. This standard of "no perceivable harm" is scientifically supportable and now applies to raw as well as processed foods. Carol Browner, Administrator of the EPA praised the law as insuring safe, healthy lives for American families.

More than 4,000 people in America die each year from salmonella in meat and poultry and as many as five million become sick. Inspection of the U.S. meat and poultry in the past has relied mostly on the out-dated senses of sight, smell and touch. Congress appropriated $574 million for major changes in inspection. Field inspectors will test for E. coli microbes, fecal contamination and salmonella. The results will be transmitted to regional offices by computer for quick analysis. We may expect more efficient and reliable inspection and a healthier public.

A bill to safeguard the nation's water supplies was signed into law by President Clinton. It requires water agencies to inform the public of bacteria and contaminants in the water. Large water systems report annually, usually in water utility bills, systems of 500 to 10,000 people may use local newspapers, and those below 500 may furnish the information on request only. A $9.6 billion fund is created for state administered grants and loans to water agencies to upgrade and repair their systems. The law also revises Federal drinking water standards.

The California South Coast Air Quality Management District announced its newest plans for reducing the smog in southern California. It is a 61 point strategy for reducing particulates and ozone in the air and is expected to ease the restrictions on businesses and motorists. No new rules such as alternative fuels for diesel trucks will be imposed to control particulates. Past studies overestimated the quantities of dust and cow manure so they will not be controlled.

Ozone producing chemicals such as oil-based cleaning solvents, paints, restaurant char broilers and several other chemicals are still targets. However, the vehicle trip-reduction measures of travel to shopping malls, schools, arenas and other destinations and car-pooling efforts--criticized as ineffective and expensive--have been eliminated. The California Air Resources Board must submit a statewide plan to the EPA in January, 1997.

The Senate finally bit the bullet and passed a bill to create an interim storage site for nuclear radioactive waste at Yucca Mt. in the Nevada Proving Grounds. The 1982 Nuclear Waste Policy Act requires the establishment of such a site by 1998. A fund set up by nuclear utilities and their rate payers to pay for the storage has already reached $6 billion. The bill must still be passed by the House and signed by the President.

1. Population Limits

Time is running out. Population control is required before many other problems in the world can be solved.

Few will disagree that there is a limit to the number of people the earth will support. Under debate is the population at that limit.

Paul Ehrlich in 1968's "The Population Bomb" [5] and with Anne Ehrlich in 1990's "The Population Explosion" [6] argues forcefully that the present earth population of 5.6 billion already exceeds the limit.

As evidence they cite massive famines, growing deserts, water shortages, dying forests, smog, holes in the atmospheric ozone layer, mountains of garbage, toxic wastes, gridlock on the highways, drugs and crime.

A measure of the population increase is the time for the population to double. That doubling time was about 1000 years from the start of agriculture in 8000 B.C. to the industrial revolution in 1650 A.D., 100 years from 1650 to 1850, 80 years from 1850 to 1930 and a minimum of 38 years from 1930 to 1968. It has since risen to the current level of 63 years.

The number of births per year is equal to the population of child bearing age times the number 0.693 divided by the doubling time. The number of deaths per year is found in a similar way and subtracted to give the change in population per year. This leads to an exponential change in the population with time. If the births per year are greater than the deaths, the population increases, if less it decreases and if equal stays unchanged.

John Bongaarts, Vice President and Director of the Research Division of the Population Council in New York City asks the question [7], "Can the Growing Human Population Feed Itself?"

Demographers project a doubling of the present population to 10 billion by 2050. Pessimists, including many environmentalists and ecologists, predict a huge disaster by then. Agriculture, natural resources and the environment, already highly stressed from current population demands, will crash, they say, from the additional population pressure.

Optimists, including many economists and agricultural scientists, are confident that continued scientific and technical innovation will enable the earth to support the 10 billion people--at a higher standard of living than currently for the developing nations and without undue destruction of the environment.

Even if the optimists are correct, this really does not avoid the catastrophe, just postpones it until a later time. Only by quickly reaching zero population growth (ZPG) can the earth's population be stabilized or decreased and the damage avoided.

An average family size of 2.1 children would keep the population constant. The goal is to quickly reach this replacement family size or less.

Nearly all of the developed countries are below replacement reproduction. Europe, US and Japan have average completed family sizes of 1.7 children. Many European nations have 1.3 to 1.5. Populations of Denmark, Austria, Italy, Germany and Hungary have reached ZPG and are slowly decreasing.

Forty-four surveys of 300,000 women over the past eight years in developing countries show birth rates have declined dramatically since the 1970s [8].

The latest contraception methods such as the pill, IUDs, and injectables are being employed. Voluntary female sterilization is a common method of family planning.

China, with a population of 1.2 billion people, reduced its birth rate rapidly with a goal of one child per family. By 1989 the completed family size was 2.4. In India the birth rate is much higher than the replacement level. To slow and stop the population growth there, contraceptives must be used by a much larger fraction of the population.

Even after the replacement level has been reached, the population keeps growing until that replacement generation dies. If the replacement level occurred today, the maximum population would not be reached until about 2050.

2. Zero Risk Pesticide Policy

In August 1996, President Clinton signed into law the Food Quality Protection Act that repealed the Delaney clause. The allowed risk of pesticides was set at one chance in a million of causing cancer in a lifetime of exposure. The new "no perceivable harm" standard is scientifically supportable.

Recently, October 1994, the Clinton administration agreed to settle a five-year legal dispute with California environmentalists and farm workers over application of the Delaney food safety clause of the Food, Drug and Cosmetic Act.

Up to 85 pesticides used by gardeners and farmers could be banned. This would be a heavy blow to backyard growers of vegetables and flowers and to farmers who grow produce and fruit and have large fields of wheat and corn.

Anyone who grows healthy roses in his garden knows that it takes more than a favorable climate. Plants require fertilizers, water and defense against their natural enemies of pests and rust. The time may come when the gardener can buy bugs to fight bugs at the corner nursery. But today we don't have that choice.

When using pesticides, it is necessary to protect people and animals and preserve the environment. However, the Delaney food safety clause carries safety to an extreme. It prohibits even a trace of pesticides in processed foods.

In the past, the Environmental Protection Agency has allowed the more sensible "negligible risk" from small amounts of pesticide. But if the settlement is approved by a US District Court in Sacramento, EPA agrees to enforce the "zero risk" in processed foods.

Can microscopic amounts of pesticides in food cause cancer? In an attempt to find the answer, typically large doses are fed to rodents; then linear extrapolations are made down toward zero doses for human exposures. No consideration is given to threshold values, below which cells recover without permanent damage.

P. H. Abelson, for many years Editor of the journal Science, now Deputy Editor, is a severe critic of the zero risk policy. In three editorials in Science [9], he led the attack on the questionable methods of risk assessment in the low-level exposures to pesticides.

Bruce Ames, a professor of molecular biology at the University of California, Berkeley, is outspoken in his criticism of current methods of determination of cancer risk. He and Lois Gold emphasize, in papers in Science [10], that cancer tests are conducted at near the maximum dose the rodents can tolerate. That causes chronic induced cell division and chronic wounding, which is a promoter of cancer in animals.

Therefore, a large fraction of all chemicals might be expected to be carcinogenic at near toxic doses. That is exactly what scientists find.

Furthermore, Ames, Gold and their colleagues suggest that the regulatory process does not take into account the vast amount of natural chemicals to which humans are exposed. Plants have natural pesticides that have enabled them to survive the evolutionary process. These chemicals are similar to the artificial ones.

Nearly all fruit and vegetables in the garden and at the market contain natural pesticides that are rodent carcinogens. Humans ingest about 10,000 times as much natural as synthetic pesticides so would be expected to have 10,000 times as much induced cancer from the plant's own pesticides as from those artificially applied.

It is sometimes wrongly assumed that humans have evolved defenses against natural chemicals but not synthetic ones. Human and animal defenses should protect equally well against synthetic and natural pesticides.

What is wrong with following the zero-risk policy and being overly safe? Much would be lost. Through the use of irrigation, fertilizers, plant breeding and pesticides, crop yields have zoomed over the last 50 years.

Pesticides have eliminated many of the catastrophes of crop failures and have removed much of the risk in farming. The cost of food has been held in check.

The abundance of food has helped the low-income families in the U.S. and abroad. Without pesticides, starvation could cause a world crisis.

Nor is zero risk the best policy for reducing cancer. The fraction of the economy that can be invested in fighting cancer is limited. That limited effort should be directed toward known highest death rate carcinogens and agents most likely to be carcinogens.

There is no debate that smoking causes more deaths than any other carcinogen. Yet some farmers are still subsidized to raise tobacco.

Few contest that fatty diets and fried food are high in carcinogens. These and many others at the top of the list, prepared by Ames and his colleagues as most likely carcinogen hazards, should be targeted.

American cancer risk policy went wrong in the past in the cases of alar and saccharine, when later experiments found them not to be the cancer risks earlier advertised. Does it make sense to continue a policy that targets the TV fear of the day?

3. Carcinogens and Anticarcinogens in Food

The new Food Quality Protection Act covers both natural and synthetic carcinogens in food. In the future both will be subjected to the "no perceivable harm" test. This is a welcome step toward a sensible carcinogen test program.

The public is continually bombarded by scare stories of cancer-causing-synthetic chemicals in food. No wonder they are worried about what to eat.

A new book "Our Stolen Future" [11] , endorsed by Vice President Al Gore, asserts that humans, animals and birds encounter serious threats from synthetic chemicals in their diets.

It's authors--Dr. Theo Colborn, senior scientist at the World Wildlife Fund, Dianne Dumanoski, reporter for the Boston Globe and Dr. John Peterson Myers, Director of the W. Alton Jones Foundation, an environmental group--claim that hormones in the diet decimate sperm counts and endocrine-disrupting pesticides cause breast and prostate cancer.

The pesticide chlorpyrifos, commonly called dursban, is used in more than 900 products tens of million times a year in homes, schools and businesses. Many people believe it is the cause of nausea, dizziness, miscarriages, birth defects, loss of memory and nerve and muscle damage.

According to a study by the Environmental Working Group and the National Campaign for Pesticide Policy Reform, 16 pesticides suspected of cancer, neurological and reproductive ailments were detected in 8 varieties of baby foods canned by Gerber, Heinz and Beech Nut.

Some Californians consider the insecticide Malathion, a nerve toxin, sprayed by helicopters to kill Mediterranean fruit flies, a threat to people as well as flies.

However, a recent study, "Carcinogens and Anticarcinogens in the Human Diet" [12], concluded that most of the natural and synthetic chemicals in the diet are at such low levels they are unlikely to pose an appreciable cancer risk.

The study was prepared by the Committee on Comparative Toxicity of Naturally Occurring Carcinogens for the National Research Council (NRC). The Committee was composed of some of the most distinguished cancer, environmental, and food scientists in the country.

It is instructive to remember that a given chemical is the same whether it occurs naturally or is produced synthetically in the lab. Table salt, sodium chloride, is a chemical made by combining one atom of sodium with one atom of chloride. If all contaminants are removed there is no way to distinguish natural from synthetic sodium chloride. The same is true of other chemical compounds.

Bruce Ames, Professor of Molecular Biology at UC Berkeley, and his co-workers have found that the fraction of naturally occurring chemicals in food, testing positive for carcinogens from rodent experiments, is not significantly different from that of synthetic chemicals. As there are many more natural than synthetic chemicals in food, he suggests that the cancer risk from natural chemicals may be much higher than from synthetic ones. And the natural chemicals are not regulated.

Among the naturally occurring chemicals that cause cancer in experimental animals are mycotoxins (chemicals produced by fungi that sometimes contaminate grains and nuts).

The NRC study finds that "diet contributes to an appreciable portion of human cancer," perhaps about a third. "Excess calories and fat appear to increase cancer risk and play a role in colon, prostate, and possibly breast cancer."

"The human diet contains both naturally-occurring and synthetic agents that may affect cancer risk. The risks associated with both depend on the level of exposure and the potency of the chemicals as well as the susceptibility of the host."

"The potencies of known naturally occurring and synthetic carcinogens present in the human diet do not differ appreciably. The human cancer risk from naturally occurring substances in the diet exceed that of synthetic dietary carcinogens."

But both levels are too low for a measurable cancer risk.

The NRC study also found that fruit and vegetables in the diet may protect against cancer. Fiber and antioxidant vitamins A, C and E appear to reduce cancer risk. And several nonnutritive agents also have shown anticarcinogenic effects.

4. The Asbestos Hazard

The billions of dollars saved by not removing the asbestos from schools and commercial buildings could be used to eradicate the more urgent hazards facing children and adults.

Not infrequently the public perception of the hazards of certain chemicals or manufactured products has been magnified by misinformation unsupported by science. Such is the case for the hazards of asbestos in schools and commercial buildings.

In 1978 Joseph Califano, Secretary of Health, Education and Welfare, in a speech quoting from an unpublished report of the National Cancer Institute and the National Institute of Environment Health Sciences said, "Five million American men and women breathe significant amounts of asbestos fibers every day." He estimated that "17 percent of all cancer deaths in the U.S. every year will be associated with previous exposure to asbestos."

This and other overstatements brought on the response against asbestos products and their manufacturers that forced Manville Corporation into bankruptcy. Manville then set up a Personal Injury Settlement Trust that is expected to pay $2.5 billion on settlements over the next 20 years.

An excellent evaluation of the hazard of asbestos [13] is given by Ralph D'Agostino, Jr., and Richard Wilson, "Asbestos: The Hazard, the Risk, and Public Policy," in the book "Phantom Risk."

At one time asbestos was woven into fire curtains for theaters and spread as fire retardant on ships. It was sprayed onto walls and ceilings in homes and commercial buildings for heat insulation, and where friction was needed, formed into objects like brake linings.

The dangers of asbestos were recognized in 1906 when 50 deaths from 1890 to 1895 in asbestos weaving mills at Calvados, France, were attributed to asbestos. In 1927 chest x-rays showed that two-thirds of asbestos workers had abnormal lungs. Lung cancer was observed in 1935 and mesothelioma of the pleura in 1956. Workers in asbestos mines, factories, weaving mills and those installing insulation have been found at risk.

Not all kinds of asbestos are equally hazardous. The most used chrysotile asbestos with curly fibers is much less dangerous than crocidolite with straight fibers. Only fibers longer than 5 micrometers (0.0002 inches) and with diameters greater than 0.15 micrometers (0.000006 inches) seem to be effective carcinogens.

Disease is caused by the dose of asbestos which is the product of the exposure and the volume of inhaled air. D'Agostino and Wilson give typical exposures.

Insulation and textile workers and asbestos miners have average exposures of tens of fibers per milliliter of air. Office worker's exposure is lower by a factor of 10,000 and school children's by a factor of 100,000.

Among the requirements of The Asbestos Hazard Emergency Response Act of 1986, schools must inspect their buildings for asbestos and take appropriate action.

The U.S. Environmental Protection Agency, EPA, allows expert assessors to determine the exposures. There is, of course, the possibility of conflict of interest as the expert assessors also remove the asbestos.

D'Agostino and Wilson estimate dosages for school children using exposures of 0.001 fibers per milliliter for 12 years of school, and of 150 days each year for 6 hours per day. They use a linear extrapolation to low exposures although there is no scientific evidence of any hazard at such low levels.

They find the lifetime chance for getting mesothelioma is about five in a million and of lung cancer less than two per million.

This is insignificant when compared to other hazards. A person riding in a car has about one chance in 70 of being killed in an auto accident during a lifetime and a pedestrian being hit by a car about 1 in 500. Risk from asbestos exposure is a factor of 10,000 less than childhood deaths of minorities that is about 1 in 20.

In most cases the exposure to people is increased by the process of removing asbestos because of the fibers that remain in the air--the lifetime dose is increased by the lingering asbestos.

D'Agostino and Wilson conclude there is no significant risk from asbestos in buildings in good condition.

5. The Alar False Alarm

The new Food Quality Protection Act replaces the Delaney clause with the standard of "no perceivable harm." This should ensure a future where "alar false alarms" can be avoided and exclusion of beneficial products from the market can be prevented.

In 1989, the CBS TV show "60 Minutes" frightened the American public with a sensational expose by Ed Bradley of the fruit ripener Alar. The backdrop displayed a huge apple with skull and crossbones.

Meryl Streep opposed Alar at press conferences and school lunches stopped serving apples.

The Natural Resources Defense Council, NRDC, a private organization, led the environmentalists to a remarkable victory. The U.S. Environmental Protection Agency conceded that Alar was a likely carcinogen and should be banned from food. The public believed Alar was a deadly toxin. The October 4, 1991 issue of Science tracks the history of this alarm.

Alar is not a pesticide. It is a chemical used on apples to regulate growth and keep them on the tree longer. When growers of Macintosh apples used alar, the apples were a deeper red, easier harvested, more uniform and economical to raise.

Unsymmetrical dimethyl hydrazine, UDMH, is a trace by-product of Alar occurring in concentrations below one percent. Pushed by the NRDC after limited toxicology studies in the 1970s showed UDMH to be carcinogenic, EPA attempted to ban Alar in food in 1985, but was stopped by its Scientific Advisory Panel who found the scientific evidence insufficient.

The 1970 studies, carried out by Bela Toth of the Eppley Institute for Research on Cancer in Omaha, Nebraska were declared deficient for quantitative risk determination by the Scientific Panel.

New studies were authorized by the EPA. In February 1989, before the rodent tests were completed, NRDC issued a report "Intolerable Risk: Pesticides in Our Children's Food", stamping Alar a "potent carcinogen," the worst threat to children's health of the 23 chemicals they studied.

NRDC concluded that one out of 4000 preschool children exposed to UDMH were likely to get cancer. Uniroyal, the manufacturer of Alar, stopped its sales because of the scare. So EPA no longer needed to make an official decision and did not publish the risk estimate of their Scientific Advisory Panel, but instead decided to wait for the final results of a rodent study.

Some toxicologists, appalled by the heavy-handed tactics, went on the offensive. Joseph Rosen, a food scientist at Rutgers University, in "Issues in Science and Technology," published by the National Academy of Sciences, blistered NRDC for using questionable math, unreliable food consumption data and dubious cancer potency.

Bruce Ames, microbiologist at the University of California, Berkeley and research scientist Lois Gold [14] affirmed that feeding rodents high doses of chemicals such as UDMH does not test whether such chemicals trigger cancer at low doses in humans.

Forced feeding itself causes "chronic wounding" that is a promoter of cancers in animals. Ames and Gold reported that "About half of all chemicals tested chronically at the MTD (Maximum Tolerated Dosage) are carcinogens (cancer producing agents)." The cell-killing effect at high doses is not the same as the carcinogenic condition that produces tumors at low doses.

In 1989, an Expert Panel appointed by the British Parliament declared that the small quantities of Alar and UDMH found in food were "no risk to health."

A United Nations panel, the same year, concluded that Alar was "not oncogenic (tumor forming) in mice" and UDMH was not a concern. The panel determined that Alar produced no cancer in mice at daily doses below 396 milligrams per kilogram of body weight and UDMH below 3.9. The group considered Alar residues in food less than 0.5 mg per kilogram not dangerous for human consumption so use of alar presented no problem.

The withdrawal of Alar from the market by Uniroyal had ended the battle won by NRDC and the environmentalists. When markets cleared out apples from fear of reprisals, apple growers lost more than $200 million. And the public was denied its healthy "apple a day."

Hopefully in the years ahead, an informed public will require proper peer review and adequate scientific tests of chemicals in food by EPA and other government bodies.

6. The Dioxin Debate

Without chlorine to disinfect our water systems, and chlorine compounds for manufacture of pharmaceuticals and products of everyday use, the health and economy of the nation could be seriously at risk.

The popular view of the American public is that dioxin is the most toxic chemical and carcinogen known. Because of its presence in Agent Orange it has been blamed for cancer in Vietnam veterans.

After the herbicide plant in Seveso, Italy, exploded in 1976, numerous cases of cancer and birth defects were predicted from the dioxin scattered around. In 1983, Missouri and the federal government purchased the entire town of Times Beach and moved its 2,000 residents across the river because one part per million of dioxin in the soil was found.

Michael Gough, in his excellent review of the dioxin debate [15], concludes, "There is no convincing evidence that it (dioxin) has caused any human disease except chloracne, a serious skin disease, and that only in highly exposed persons."

Dioxin has never been produced commercially. It is a by-product of the production of 2,4,5-trichlorophenol that was used in the manufacture, now discontinued, of bactericides and herbicides. It is also present in the smoke of burning wood.

Philip Abelson in an editorial [16] in the journal Science points out that wood smoke contains more than 100 organochlorine compounds. Included is the extremely toxic 2,3,7,8 tetrachlorinated dioxin.

He gives estimates that Canadian forest and brush fires inject 10 times as much dioxins into the atmosphere each year as the Seveso plant accident. And 2,3,7,8, tetrachlorinated dioxins from biomass burning by developing nations is 10 times the U.S. Environmental Protection Agency estimates of current American emissions caused by man.

The EPA uses a linear model with no threshold to assess the safety of dioxins. As a consequence, the permissible level of background dioxins is a factor of 150 to 1500 lower than those for Canada and Europe.

Dioxins have entered the food chain and are present in fatty tissues of our bodies. Half leaves the body in 7 to 20 years. They are found even in tissues of 70 year old men who have spent their lives in the desert.

Early Swedish studies of people with low level exposures to dioxin gave indications of increased soft tissue sarcomas. However, improved follow-up investigations by Swedish scientists and by scientists from the United States and from New Zealand gave no statistically significant confirmation.

Workers in chemical plants that formerly manufactured dioxin laden chemicals have dioxin levels 100 to 1000 times those in other industrial plants. Studies of workers exposed to high levels of dioxin, according to Gough, "do not support the conclusion that dioxin has been a major cause of cancer in humans. In fact, those studies cast doubt on whether dioxin has caused human cancer at all."

Children exposed to the Seveso herbicide plant explosion who developed chloracne had an average dose of dioxin, pound for pound, that was enough to kill half of all guinea pigs so exposed. But after 10 years none of the 193 chloracne children have died of cancer. The overall cancer rate in the total exposed population was lower than in the unexposed control population. A tumor registry will enable these studies to be continued indefinitely.

The average dioxin level in Vietnam veterans who sprayed Agent Orange in Vietnam is three times the level in other Vietnam veterans and veterans who never served in Vietnam. In a critical review in 1988, of Agent Orange in Vietnam, Brian MacMahon, former chair of the Department of Epidemiology at the Harvard School of Public Health concluded, "there is no reasonable basis" to decide that Agent Orange caused "increased risk of STS (soft-tissue sarcomas) or NHL (non Hodgkin's lymphoma)."

Efforts were initiated by EPA administrator Reilly and his staff in 1991 to relax the permissible levels of background dioxin and other chemicals. Earlier efforts had failed. Instead, goaded by Greenpeace and its allies, EPA might even curtail or ban the production of chlorine and its compounds.

7. Credibility of Asteroid and Air Pollution Risks

It is clear that the uncertainties in the APCD cancer risk values are immense. In fact, if dose thresholds occur, the risk of cancer from current exposures of particles in the atmosphere could approach zero.

Risks of death to people by asteroids striking the earth and from cancer by air pollution have little in common, except claims in the media that they are about equal.

The May 1996 issue [17] of the magazine Discover gives the statistic, "The risk of your being wiped out (with nearly everybody else on the Earth) next year by a catastrophic comet, meteor, or asteroid impact: 1 in 20,000."

Depending on your worry index, the asteroid risk may be cause of concern. It is comparable to other risks we face; one-third the risk of being killed in an automobile wreck and 10 times the risk of being killed in a commercial airplane accident.

The risks of autos and planes are well known from many years of past experience. They are based on about 40,000 auto deaths per year in the United States and tens of commercial airplane deaths. To calculate the risk for autos just divide the number of auto deaths by the number of auto riders (the U.S. population), about 240 million, to obtain the risk: 1 in 6,000.

How was the calculation made for asteroids (comets and meteors are combined here with the asteroids)? Man's recorded history goes back only about 5,000 years. During that time there's no example of an asteroid striking the earth big enough to cause tidal waves, or change the weather or climate enough to cause even a few deaths, certainly not nearly everyone on the Earth.

The evidence we do have is the demise of the dinosaurs 60 million years ago. Most scientists agree that a comet or asteroid about 15 kilometers in diameter crashed just off the Yucatan Peninsula of Mexico. A huge cloud of dust and water was kicked up into the atmosphere. For years, light from the sun was absorbed by the cloud. The dinosaurs died out in a few years from lack of food or a climate change.

Such a catastrophe might or might not cause the death of part or all of the human population on the earth. Assuming no one survives, an estimate of the risk would be 1 in 60 million per person per year. This risk is a factor of 3,000 less than the risk quoted by Discover magazine.

So where did Discover get its value of the risk? The distribution of the diameters of asteroids in space has not been measured. However, some scientists think that the number should decrease as one divided by the square of the diameters of the asteroids. The risk from an asteroid with a diameter of 200 meters would be 1 in 20,000 for each person each year as quoted by Discover--if everyone on earth were killed.

For the 200 meter asteroid, it seems unlikely that all people on earth would die. A tidal wave from an asteroid landing in the ocean might wipe out some coastal cities. Certainly cities hundreds of meters above sea level would not be destroyed. We don't know the extent of damage that would be caused by the water and dirt parked in the atmosphere but there is no reason to believe all or even a sizable fraction of the population would die. If only one-tenth dies, the risk would be reduced to 1 in 200,000 per person per year; if one-hundredth dies, the risk would be 1 in 2 million per person per year.

The risk of getting cancer from toxic air pollutants in downtown Santa Barbara is given by the County Air Pollution Control District as 1 in 2,000 for each person in a lifetime. That risk assumes a lifetime of 70 years spent downtown for 24 hours every day. Dividing by 70 gives the risk for each person each year: 1 in 140,000. This risk is lower than the Discover Asteroid risk by a factor of 7.

The APCD stresses that cancer is not caused by ozone in the air. In high enough concentrations ozone can be detrimental to health. It can cause asthma, bronchitis, eye irritation, nausea and headaches.

Cancer in humans may be caused by particulate matter in the air such as dioxins, chromium VI, cadmium, arsenic, benzo[a]pyrene (BaP), nickel and other compounds. The risk of cancer depends on the toxicity of the particulate matter and the magnitude of the exposure.

It seems reasonable to inquire how the APCD calculates the cancer risk. An example is given in the report on "Benzo[a]pyrene as a Toxic Air Contaminant" prepared by the California Air Resources Board and the Office of Environmental Health Hazard Assessment for the California Environmental Protection Agency, December 1993.

Since no adequate studies of the carcinogenicity of BaP to humans have been made, the toxicity value is estimated from rodent experiments. To induce cancers in the animals, typically large (near lethal) doses are administered. To obtain the toxicity for humans at levels of BaP found in the air, the rodent data are then linearly extrapolated downward by a factor of 100,000. No consideration is given to the likely possibility that threshold dose values exist, below which cells recover without permanent damage.

8. The Endangered Species Act

A fair law is needed that protects the rights of citizens as well as the habitats of endangered species. A possible model is the plan initiated for the 37,000 acre wildlife preserve in Orange County.

The Endangered Species Act, is again before Congress. A bitter battle is expected with a decision on its revision and extension before the end of 1996.

The act was passed in late 1973. Its stated purpose is "to provide a means whereby the ecosystems upon which endangered species and threatened species depend many be conserved."

The Committee on "Scientific Issues in the Endangered Species Act," appointed by the National Research Council and chaired by Michael T. Clegg of the University of California, Riverside, concluded [18] that "The Endangered Species Act is based on sound scientific principles."

The heart of the law is habitat protection. The Committee agreed that habitat conservation and recovery plans are essential to any program to protect endangered species. It was not asked to comment on social and political consequences of the Act's usage.

Much of the planning and management of conservation efforts falls on the U.S. Fish and Wildlife Service. That agency and others must make decisions often based on insufficient data, uncertain predictions, conflicting management objectives and disagreements over courses of action.

There are never sufficient funds to carry out the necessary scientific studies, plans and oversight. No wonder their success has been fragmentary at best. A measure of this success is the recovery of species from endangered or threatened status.

The wildlife service reported to Congress in 1992 that 10 percent of the listed endangered or threatened species improved, 28 percent was stable, 33 percent declined, 2 percent became extinct and 27 percent was unknown. Only six species have been delisted since 1973 because of successful recovery: the Palau dove, owl and flycatcher; the Atlantic coast brown pelican; the gray whale and the Rydberg milk-vetch.

The act would be workable if the public's only goal were to protect endangered and threatened species and their habitats, Fish and Wildlife Service and other required state and local organizations were properly staffed and funded and private property owners were properly compensated for their forced participation.

But proper staffing and funding is very expensive and there are many other economic, political and social considerations and rights that are placed in jeopardy by the Act.

We all know horror stories like that of Yshmael Garcia in Riverside County who blames the fire that burned down his home on protection of the endangered Stephen's kangaroo rat. He was not permitted to clear away the brush around his home.

Then there were the criminal charges brought against Kern County farmer Taung Ming-Liu who was accused of killing five kangaroo rats on his farm. The charges were ultimately dropped.

Five generations of Domenigoni family raised grain on their farm in western Riverside County. They left a different one-quarter of their land fallow each year. Kangaroo rats invaded and they were prohibited from plowing the 800 acres for three years until the rats moved out. This cost the

Domenigonis $400,000 in lost crops and fees for attorneys and a biologist. After six years of study of kangaroo rat recovery plans, much of Riverside County property still remains in land-use moratoriums.

The U.S. Supreme Court ruled in June that the federal government can stop landowners from cutting trees and damaging the habitat of the northern spotted owl and other endangered species. Under President Clinton's compromise plan, logging levels would be limited to about one-quarter of the yearly average during the 1980s. This satisfies neither the environmentalists nor the logging industry. At stake are thousands of jobs and billions of dollars.

The gnatcatcher inhabits coastal sage scrub in southern California. Around 2,000 pairs are thought to live on about 100,000 acres. If declared endangered, one study finds 200,000 jobs eliminated in Riverside, San Diego and Orange counties and economic losses up to $16 billion.

Although still in the planning stages, developers are working with federal, state and local governments to establish regional sanctuaries for the gnatcatcher and other species. Other ideas are compensation for restrictions on use of private property, tax rebates and tradable credits in endangered species habitat.

9. Telescopes and Red Squirrels

The biggest losers of all may be the environmental organizations who are forfeiting their credibility with a growing fraction of the public. Fortunately, the Congress has an opportunity to rectify the inequities in the Endangered Species Act.

In California, where we are often immobilized by battles between environmentalists and other citizens over threatened or endangered species, it is comforting to learn that we are not alone.

While it is the kangaroo rat in Riverside county, the California gnatcatcher in San Diego county and the spotted owl in northern California, Arizona has its own problem.

Bruce Walsh, Roger Angel and Peter Strittmatter, faculty of the University of Arizona, report in the journal Nature [19] how red squirrels have endangered telescopes. It's not that the squirrels are hiding nuts in the mirrors or competing for telescope viewing time. It's the use of the endangered red squirrel to oppose the Mount Graham Observatory.

In the early 1980s, the University of Arizona and the Smithsonian Institution, after an extensive search, selected a site at 10,500 feet on Mount Graham as an ideal spot for future telescope observations. The site possessed clear, dark skies, low humidity and easy access. Since it already had a paved 30 mile, 2 lane road, transmitter stations, 93 homes, an artificial lake and 250,000 visitors a year, any impact on the environment would be small.

The U.S. Forest Service carried out a 4 year evaluation of biological and other environmental impacts. Red squirrels were classified as an endangered species. A biological opinion, under provisions of the Endangered Species Act, permitted 3 telescopes limited to 8.6 acres, with several protective provisions that included foresting 60 acres of road, creation of a red squirrel refuge and a squirrel-monitoring program with full time biologists, at observatory cost.

The observatory was ratified in the 1988 Arizona Idaho Conservation Act that adopted results of two environmental impact studies and the biological opinion. Two telescopes have been constructed and are in use.

The third, the Large Binocular Telescope, LBT, with two 28 ft diameter mirrors on a single mounting and with new adaptive optics to remove atmospheric aberrations will be more sensitive than any other existing optical telescope.

It is expected to extend the observational limits and discover structure in the universe at times earlier than any before.

Plans for the LBT have been modified in many ways to address public concerns. The project has an 'approval rating' of more than 65 percent in Arizona with less than 20 percent opposed and approval by more than 90 percent in Graham County. Yet the LBT has become the main target of activist opponents in the Maricopa Audubon Society.

The new lawsuit by environmentalists against the Forest Service claims the agency had no authority to revise the LBT site slightly to a place with less effect on the endangered red squirrels. The suit suggests an agenda different than protection of endangered species.

The red squirrel is one of the most numerous mammals in the United States. It is sometimes threatened by inadequate food when the cone crop fails, and by excessive logging. But few squirrels will be affected by the observatory that includes only 8.6 acres out of a habitat of 11,000 acres.

The biological opinion estimated conservatively that construction of the observatory would reduce the squirrel population by less than one. Opponents claim concern about observatory influence on red squirrel behavior, but 25 biologist years of monitoring found no measurable effects, even in the immediate vicinity of construction.

The telescope should be built at the highest possible altitude for the best observations. This is at the spruce-fir elevation which the opponents claim is critical to the red squirrel habitat. The biologists have found, however, that the main squirrel population lives outside the spruce-fir zone.

To quote the University of Arizona scientists, "Clearly, the environmental impact of the observatory has been exaggerated, and the measures to protect the squirrel distorted, to halt the project.....It is one thing to use the (Endangered Species) Act to protect a truly endangered species from a real threat. It is quite another to use it to stop projects that have no significant environmental impact."

The loss of time and expenses of lawyers, professional protesters, and protection of the telescope from threatened damage is borne by federal agencies and the University of Arizona--both supported by public taxes--and by private contributions to environmental organizations.

Unfortunately, the endangered red squirrel is not the first misuse of the Endangered Species Act nor is it likely to be the last. And the endangered species are not restricted to red squirrels or kangaroo rats but include astronomers wishing to build the best telescopes for exploring the universe and young men and women who want new homes and need new jobs.

10. Low Level Radioactive Waste

Depositing low level radioactive waste at the Ward Valley site will protect present citizens of California and future generations from possible threats to their health.

The National Academy of Science's 17 member panel recently released its report [20] on the safety of the Ward Valley Low Level Radioactive Waste Depository.

The panel concluded that the Ward Valley dump near Needles in the Mojave Desert would not contaminate the Colorado River, 20 miles away. Nor would it pose a health risk to the millions of people in Southern California drinking Colorado River water.

The panel also recommended that more tests were advisable to determine whether small amounts of waste could seep into the water table below the dump. Fifteen agreed that this need not hold up construction of the repository.

Critics were also concerned about possible plutonium-239 that might leak into the Colorado River. The panel acknowledged that small amounts might possibly reach the Colorado after 100s of 1000s of years but would be negligible compared to present natural levels and would meet accepted regulatory health standards.

Interior Secretary Bruce Babbitt had requested the study before he would approve the transfer of the federal land for the Ward Valley site to California.

Two weeks after completion of the study, Governor Pete Wilson released a letter to Babbitt urging him to sign the transfer and implement the recommendations of the panel.

Radioactive isotopes trace chemical reactions and their rates of change in time, and migration to different places. They are indispensable to research experiments with matter, animals, plants and humans.

Hospitals use the radioactive isotopes to diagnose illnesses. Tests are often carried out in diagnostic labs.

Some isotopes like technetium-99m, which has a half-life of 6 hours, can be kept on the shelf for a few days then discarded safely with the normal waste. The half-life is the time for one-half the atoms of the isotope to decay.

Others with longer half-lives like tritium, cesium-137 and carbon-14, with half-lives of 12, 30 and 5,730 years, should be isolated in a Low Level Radioactive Waste (LLRW) dump to protect the public.

The problem of disposal of LLRW by universities, private research labs, hospitals and diagnostic labs has been festering since the passage of the Federal LLRW Policy Act of 1980 that requires each state to designate a site for these wastes.

The Ward Valley site selection process began in 1984. Studies were carried out but decisions were not made. California institutions continued to ship LLRW to Nevada and Washington. On January 1, 1993, that was stopped by a new federal law.

For a while California LLRW was then shipped to a site at Barnwell, S.C., at much higher cost, but that too was prohibited in 1994. There is no place for the wastes to go. Radioactive waste is temporarily stored at research universities like UCSB, research hospitals like those at UCSF and UCLA and other labs and hospitals across the state.

Senator Barbara Boxer, D-Calif., has been an outspoken critic of the Ward Valley site. In its place she has proposed above ground tanks to hold radioactive isotopes temporarily until they decay away. The isotopes would then go to the normal garbage dump.

It is difficult to take this proposal seriously. It is little different than the current dilemma. Would each generator of radioisotopes store its own? How does this solve the problem of the much used isotopes with half-lives of decades to thousands of years?

In case of fire, earthquake, carelessness or theft is it really preferable to store the radioactive isotopes in Los Angeles, San Francisco or Santa Barbara than in the Mojove Desert? Would the drinking water of Californians be better protected by the temporary holding tanks in these cities?

11. The Yucca Mountain Nuclear Waste Repository

The Senate passed a bill in August, 1996 to create an interim storage site for nuclear radioactive waste at Yucca Mt. The public, House and President now have the opportunity to support the Senate in the selection of the Yucca Mountain site.

Perceptions of the public too often are molded by renegade scientists abetted by reporters who offer controversy rather than an informed choice. An example is the Yucca Mountain Nuclear Waste Repository.

The Nuclear Waste Policy Act was passed by Congress in 1982 to establish a national repository for safe, 10,000 year or longer storage of spent nuclear reactor fuel assemblies.

The Department of Energy (DOE) was given the responsibility for selection and development of the site. The obvious choice was Yucca Mountain in the southwestern corner of the Nevada Test Site near Death Valley.

The fuel assemblies could be encased in glass, enclosed in stainless steel cylinders and sealed in tunnels in the mountain under 1500 feet of earth. The rainfall is only 6 inches per year and the water table, 1,000 feet below, is protected by layers of tuft.

Hundreds of wells in the ground and tunnels in the mountains were earlier drilled for weapons tests before the underground test ban. The geology and hydrology of the region is well understood.

However, the development plans received a setback in 1989 when a report by Jerry Szymanski, a hydrologist on the DOE staff, suggested that groundwater in the past 10,000 years had risen several times to the level of the proposed repository.

He claimed that earthquakes could cause "seismic pumping" that would repeat the rises in the future. Radioactive isotopes might get into the water table and migrate to regions away from the repository. The New York Times Magazine on Nov. 18, 1990 featured the story, "A Mountain of Trouble" by William Broad.

To resolve the matter, the National Academy of Science in 1990 established [22] a panel of experts to evaluate the report and make recommendations. The panel concluded its report, "Ground Water at Yucca Mountain, How High Can it Rise?" in April 1992.

The panel found that "None of the evidence cited as proof of ground-water upwelling in and around Yucca Mountain could be reasonably attributed to that process (seismic pumping)."

And the features ascribed to the rising water level were caused by volcanic processes that occurred 10 to 13 million years ago. The features contained contradictions that made the upwelling ground-water origin geologically impossible. They were classic examples of arid soil characteristics recognized world-wide.

More recently the same reporter on the front page of the New York Times [20], stirred a new controversy with a story headlined "Scientists Fear Atomic Explosion of Buried Waste; Debate by Researchers; Argument Strikes New Blow Against a Proposal for Repository in Nevada."

Charles Bowman and Francesco Venneri, scientists at Los Alamos National Laboratory (LANL), had suggested that plutonium in the spent fuel rods at the Yucca Mountain waste repository could leak from the stored containers into the adjacent rocks. Moderated fission neutrons in the plutonium could cause an explosion and that would set off additional explosions in surrounding containers.

LANL immediately appointed 3 internal review committees--a blue team to duplicate the Bowman-Venneri results, a red team to find its errors and a white team of neutral senior scientists to make recommendations to management.

The blue team could not validate the Bowman-Venneri calculations for extracting the plutonium from the spent fuel and concentrating it in a critical mass that would explode. James Mercer-Smith, chair of the blue team, said the Bowman-Venneri calculations were just wrong.

Art Forster, chair of the red team, said the most important calculation to determine whether the explosion would occur was neither carried out nor estimated. The white team concluded that the probability was essentially zero that the Bowman-Venneri explosion could occur.

Most outside reviewers have been equally critical. An additional important test will come when the Bowman-Venneri paper is peer reviewed on submission for publication. To date, September 1997, such a peer reviewed paper has not been published.

12. Transport of Spent Nuclear Reactor Fuel

The Santa Barbara City Council and later the Santa Barbara County Board of Supervisors passed resolutions opposing the transport of spent nuclear reactor fuel through the city and county, respectively. At the county hearing about 25 presentations were made in favor of the resolution and one against. My presentation against was the only one giving results of calculations of the risk to the residents of the county. That presentation was continually interrupted by boos and catcalls. Do we need a better demonstration of the need for science education in this country?

On October 10, 1995 the Santa Barbara City Council passed resolution 95-144 opposing the transportation of spent nuclear reactor fuel and high-level radioactive waste through Santa Barbara and other heavily populated areas. A News Press editorial was in favor of the resolution. Letters to the editor agreed.

A study, "High Level Radioactive Waste Shipments by State and by Method 1998-2030" [21] by the Nevada Agency for Nuclear Projects, estimates that about 40 casks per year of spent nuclear reactor fuel will be generated in California over the next 32 years. About 38 would be shipped by rail and two by truck from a pair of nuclear reactors at Diablo Canyon near Avila Beach and from the three reactors at San Onofre at San Clemente.

These casks would be hauled about 600 miles to the designated high radioactivity waste repository at Yucca Mountain, Nevada.

The City Council was rightfully concerned about accidents that could possibly spread radioactivity and endanger the public.

Yet in examining the City Clerk's file of back-up information for the resolution, minutes of City Council meetings and staff memos, there appeared to be no reports or references to reports that address the dangers of transporting spent nuclear reactor fuel.

Fortunately, this void is filled by a March, 1987 report of the U.S. Nuclear Regulatory Commission, "Transporting Spent Fuel--Protection Provided Against Severe Highway and Railroad Accidents" authored by William R. Lahs [22].

He points out that the rail accident rate is about one per million miles and the truck rate about six per million miles. However, the public is protected because the cask design must be certified by the Nuclear Regulatory Commission against impact, puncture, fire and water immersion.

The cask must withstand a 30 foot drop onto a flat horizontal unyielding surface. It must be resistant to a temperature of 1475 degrees Fahrenheit for 30 minutes, and not leak if put under 50 feet of water for 8 hours. A cask that survives these tests will not be penetrated in 994 of 1000 accidents. It also includes a layer of lead sufficient to keep radiation levels outside the cask below compliance levels.

In four out of 1000 accidents, there will be leakage of radioactive materials above a negligible amount. In two out of 1000, the radiation could exceed compliance values by a factor of three and in one out of 1000, by a factor of 30.

To calculate the risk for Californians, a reasonable assumption is made that each train transports 10 casks and each truck one. This results in four trips by rail each year to Yucca Mountain and two trips by truck, for a total of 2400 miles traveled by rail and 1200 miles by truck.

Thus the number of rail accidents in California giving radiation above compliance is one in 100,000 years, and of truck accidents, three in 100,000 years.

If the danger is confined to the vicinity of Santa Barbara alone, a travel distance of 30 miles along the railroad and highway 101, casks would come only from Diablo Canyon. So the risk is reduced by another factor of 40. The risk is then about one accident giving radiation above compliance by rail in four million years and one by a truck in a million years.

Some people worry about earthquakes. A very conservative estimate is taken that one large earthquake every 10 years, of magnitude greater than 6.5, will occur along the rail or highway corridor, 30 miles wide, to Yucca Mountain. Suppose the trip travel time is 24 hours. This gives the chance of earthquake during one trip of about one in 3600.

Suppose an accident occurs if the earthquake center is within 15 miles of the train or truck. The chance of an accident caused by an earthquake is then 4 in 10,000 years, a very conservative estimate. The temblor's shock occurs over seconds or minutes so should be less violent than those described above. Therefore the casks should suffer less damage. These conservative estimates for accidents during earthquakes in California, giving radiation above compliance, are 1 and 0.5 per 10 million years for rail and truck.

The Lahs report concludes that the overall annual risk from transportation accidents, involving spent fuel shipments in America, should be less than about 0.0001 latent cancer fatalities per year. In the 18 years since 1977, thousands of shipments of commercial spent nuclear reactor fuel have been made without adverse radiological consequences to the public.

This compares to 400,000 deaths that actually occur each year in America from cancer due to other causes.

CHAPTER V

ENERGY SCIENCE

Perhaps the most important scientific decision the American public must make in the coming years is the choice of the source(s) of energy to produce the electricity we use in our homes, businesses and industries.

We are all aware of our dependence on electricity. In our homes we rely on lights, furnaces and air conditioners; in the kitchens on clocks, timers, ovens, refrigerators, freezers, dish washers, mixers and garbage disposals. For cleaning our carpets we need sweepers and for our clothes, washers and dryers. Our entertainment centers contain radios, stereos, tape players, disc players, TVs, videos, and VCRs. In the study we require typewriters, computers, printers and modems. For our garden we use hedge trimmers, lawn edgers and chain saws and in the shop--drills and circular and band saws. And our homes are protected with alarm systems and garage door openers. All run on electricity.

No wonder we are paralyzed when the electric power goes off. And the problems with businesses and industries can be even more severe.

The fossil fuels of coal, oil and gas have long been major sources for the generation of electricity. Gasoline and other petroleum fuels have propelled our vehicles. This is no accident. They competed with other sources in the 19th and early 20th centuries and were the clear winners. Their energy densities, energy divided by mass, are high. The fuels are portable--can be delivered by truck or train. They satisfy the needs of people for heat, electricity and propulsion.

Contrary to claims of some detractors, there is no conspiracy among industrial giants, or industry and government to force fossil fuels on the public. It is equally fallacious to suggest that big fossil fuel producers or users have banded together to block the growth of other sources of electrical power. There is plenty of talent and venture capital available for new ideas, products and energy sources that have merit. The computer and biotechnology industries are good examples. The coal, oil and gas companies will be the first in the future, as they have been in the past, to try out and develop worthy sources of power.

DEREGULATION OF ELECTRIC POWER

A lot is expected of our electric utilities. Usually the public is well served, in most communities interruptions occur infrequently. However the blackout on Aug. 10, 1996 was massive. Because of hot weather in Oregon, a power line sagged into a tree at 2:06 pm and lost power. It was followed by a second at 2:56 pm and two more at 3:47 pm. An 860-megawatt power generator shut down. Wild line power fluctuations occurred. One minute later main lines from the Pacific Northwest to California lost power removing 4,300 megawatts from the grid. Four million people in nine western states were left without power for many hours.

Several problems existed, none individually critical, that collectively caused the outage. Among these, additional power, normally present, was not available. About 20% of the power along the Columbia River in Oregon was out of service for maintenance. Some of the water from the Dalles Dam power plant on the Columbia River had been diverted to assist the running of the Chinook salmon so the dam was generating less power than customary. The plant, like others, is subject to the Endangered Species Act. Randy Hardy, Chief Executive of the Bonneville Power Administration, the agency responsible for the Oregon transmission lines, testified that the decision to curtail power in the Dalles plant saved about half a dozen endangered fish. Blackouts of this kind are clearly preventable.

In April, 1996 the Federal Energy Regulatory Commission mandated the opening of the utility transmission lines to outside wholesale energy providers. Markets will be opened nationwide, new competitors are expected and new services will be created. It is hoped that industry and consumers will be saved billions of dollars every year.

California is leading the way in the deregulation of electrical power. The deregulation bill, AB 1890, was passed and signed in the Fall of 1996. Power companies from all over the nation will be able to compete to provide electrical power. The current utility providers will be reimbursed about $27 billion for assets such as power plants and alternative energy purchase commitments. These costs will be paid by the customers in a surcharge added to the monthly bills from 1998 to at least 2002. Residential customers are promised a 10% rate cut in 1998, one welcomed by California residents who customarily pay rates 50 percent higher than the national average.

ALTERNATIVE ENERGY MYTHS

Alternative energies that include geothermal, wind power, biomass, solar collectors and photovoltaics, for the last 25 years have been ballyhooed as the energies of the future. We are told by their advocates that alternative energies are renewable, nonpolluting and cheap. Unfortunately production of electricity by alternative energies has not lived up to the hype.

The Geysers in California, the showcase of geothermal energy, has demonstrated that geothermal energy is not inexhaustible. Production peaked in 1988 and has since been steadily going downward. California taxpayers were stuck with costs of state-built unused and early-retired electricity generating plants when the quantity of steam and hot water generated was lower than predicted. Once the present hot water and steam is removed it may take hundreds of years to restore.

Hillsides covered with wind power machines in windy areas are both sight and sound polluters. Noise levels in their vicinity are excessive, approaching those of large busy airports. Tens of square miles of wind machines in mountain passes do not beautify the landscape. Even worse are the abandoned fields littered with fallen wind machines and parts scattered around on the ground.

Biomass, when burned to produce energy, is a greater polluter than the fossil fuels of coal, oil and gas. Biomass production also requires large land areas. Already competition for land for homes and businesses, farming, forests, recreation, parks, hunting, grazing, mining, wilderness and habitat for endangered species is one of the biggest problems facing the federal government.

The sun's energy striking each square foot of surface of the earth is low. It is blocked by the earth at night; and at high latitudes the sun's rays strike the earth at large angles to the vertical, reducing the energy striking each area. Clouds form and storms occur that reduce the energy striking the earth. Therefore, large areas of collectors, several square miles in area, are required to furnish the energy produced by one large coal, oil, gas or nuclear reactor plant. The sun's radiation is typically reflected to central towers where the heat is stored in water or salt, then used to generate electricity. Tens of square miles of glass or metal reflectors create the sight pollution. And competition for land use, just as for biomass and wind power machines, raises very difficult questions.

Photovoltaic cells have the advantage of being able to convert light from the sun directly into electrical power. Because they are more efficient than solar collectors, they need less collection area. At the present time commercial cells are much too expensive to replace fossil fuels and nuclear reactors. However, they are quite useful in remote locations where cost is not an issue. With further research it is possible that the cost of photovoltaic cells may eventually become competitive.

Most of the public is unaware that alternative energies furnish a negligible fraction of the electrical power in the United States. This has been true in the past and will likely be true for the foreseeable future. In 1993 [1] coal produced 57 percent, nuclear reactors 21, oil 3.5, gas 9.0, hydroelectric 9.2 percent and all alternative energies combined less than about 0.3 percent. Geothermal energy produced most of the alternative energy with wind machines, solar collectors and solar photovoltaics each less than 0.1 percent.

Even with the state and federal regulations supporting alternative energies with tax incentives and guaranteed favorable rates by utilities for all energy produced, the alternative market share has been about that same low 0.3 percent over the last 20 years. In the free market of the future, which will remove favorable treatment for the alternatives, they will contribute yet a smaller percentage of electrical power to the nation.

In the spring of 1996, wind power was dealt a major blow when Kenetech Windpower, Inc., the largest wind farm developer in the United States, filed for bankruptcy protection under Chapter 11 of the U.S. bankruptcy code. Reasons cited were that the price of electricity had dropped in the last five years; Kenetech wind turbines had developed some mechanical problems; Kenetech was involved in a series of lawsuits over warranty claims; and they predicted that tax incentives would be terminated in the future and were concerned about the recent drop of private and public investment in wind technology.

Before the invention of the steam engine in the 1700s, wind power was used to pump water, grind grain, and perform other early industrial tasks. Quaint windmills dotted the countryside in Holland and other European countries. In the 19th and early 20th centuries nearly every farm in America had a windmill to pump water for livestock and for use in the home. However, times have changed. Because windmills are not reliable--the wind is variable--many have been replaced by natural gas or electrical power.

A solar collector plant, Solar II, near Barstow, Calif. in the Mojave desert went on line in June 1996. It was advertised as the most advanced in the world. At 10 megawatts it produces about one percent as much electrical power as a coal or nuclear reactor plant. Run by Southern California Edison Co., it reflects sunlight from about 2,000 mirrors to a central tower. Molten salt stores the heat that is transferred through a heat exchanger to produce steam. The steam runs a generator that produces the electricity. Solar II, not unexpectedly, has had its growing pains. It was shut down for an extended period after turn-on in the summer to correct problems with hot spots in the molten salt that ruptured its container.

Nuclear Energy

The first half of the 20th century was the nuclear age. In 1911, Ernest Rutherford, an English Nobel Laureate in physics, and his coworkers founded nuclear physics when they discovered the hard dense core of the nucleus of the atom. The neutral hydrogen atom has a proton in the nucleus and an electron in orbit around the nucleus. The other particle in heavier nuclei, the neutron, was discovered by another English physicist, James Chadwick. The heaviest naturally occurring nucleus is uranium-238 made up of 92 protons and 146 neutrons.

When uranium-235 or plutonium-239 captures a neutron it may split into two or more lighter fragments, releasing about a million times as much energy as from a chemical reaction such as the burning of hydrogen and oxygen. This is called fission. German physicists, Otto Hahn and F. Strasmann discovered fission in 1939. Many scientists around the world immediately recognized that controlled fission would be an abundant source of energy that could be used for generating electricity; and that run-away fission could be used for huge bombs.

Italian physicist, Enrico Fermi, immigrated to the United States and in two years, with a team of scientists working under the stadium at Stagg Field at the University of Chicago, built a nuclear fission reactor. The future of nuclear energy looked bright.

Nuclear power is so attractive because of the high energy density in nuclear fuel. Each kilogram of nuclear fuel produces about one million times the energy of one kilogram of chemical fuel (one kilogram equal 2.2 pounds). One kilogram of uranium-235 generates about the same amount of electrical energy as one million kilograms of coal, oil or gas. Therefore fossil fuels have about one million times as much waste as nuclear fission for the same amount of energy. Furthermore, much of the waste of fossil fuels goes into the atmosphere where it causes a significant fraction of the pollution of the environment. Most of the rest of the pollution comes from burning gasoline and other fossil fuels in cars and trucks. The used nuclear reactor fuel rod waste is contained within the reactor and, if it were not for the anti-nuclear activists, could be safely hauled away to secure underground storage facilities like Yucca Mountain.

Because of the opposition to nuclear reactors, no new permits have been issued in the last 20 years. Production of electricity by nuclear reactors has leveled off at about 22 percent. The activists' claims that nuclear reactors are dangerous is false. Not one person has been killed by commercial nuclear reactors operating in the United States over the 40 years or so that they have been producing electricity. No other energy source has such an excellent safety record.

The anti-nuclear reactor activists cite the Three Mile Island Accident as proof that nuclear reactors are unsafe. On the contrary, it proves nuclear power plants are safe. In that accident not one person was killed or injured. The small amount of radiation that escaped was hardly measurable above the earth's natural cosmic ray and earth background radiation.

President Carter's Commission on the Three Mile Island Accident, chaired by John Kemeny, President of Dartmouth College, reported the findings of its study in "The Need for Change: The Legacy of TMI." In summary it said, "Based on our investigation of the health effects of the accident, we conclude that in spite of serious damage to the plant, most of the radiation was contained and the actual release will have a negligible effect on the physical health of individuals......It is entirely possible that not a single extra cancer death will result." The commission included 12 distinguished members, one of whom was the then Governor of Arizona, now Secretary of the Interior, Bruce Babbitt.

Studies of the incidence of cancer in the area surrounding the reactor have found no statistically significant evidence for increased cancer due to the accident [2].

But the critics ask, if not Three Mile Island then how about Chernobyl? There is no excuse for the melt down at Chernobyl. That Russian reactor was an unsafe design manned by careless operators with slipshod practices. American designs and others around the world based on American designs are safe. Since the Three Mile Island accident nearly 30 years ago, US reactors have been improved with newer designs, instruments and practices. Chernobyl will never happen in the United States.

Nuclear reactors have received a bum rap. My perception of the reason is that the public does not understand the fundamental differences between nuclear reactors and nuclear bombs. Because of this misunderstanding the public has suffered from incorrect energy policies. The United States has not taken advantage of its early world lead in safe, clean, cheap nuclear energy.

A nuclear reactor uses controlled neutron fission of uranium or plutonium to reach steady energy production. A nuclear bomb uses run-away neutron fission where all the energy is released in a fraction of a second. The designs of a reactor and a bomb are entirely different. A fission nuclear reactor cannot explode as a bomb and a bomb cannot generate steady electrical energy.

The analogy to chemical reactions may help. Electrical power is produced by the controlled burning of coal, oil and gas. The rates of the chemical reactions of carbon and hydrogen with oxygen that release energy are controlled at the power plants. On the other hand chemical reactions of certain chemicals like dynamite and TNT when ignited release all their energy in run-away chemical reactions in a very short time. They are chemical bombs. We know how to control chemical burning in stoves, ovens and furnaces. Therefore we don't outlaw them from our homes because we're afraid they'll blow up like dynamite bombs.

Of course, we're all concerned about the proliferation of nuclear bombs. We must do everything in our power to reduce their possible use. But that doesn't mean that we must throw out the baby with the bath. Nuclear energy is too important to our lives to casually toss aside because of the criticisms of some well organized activists--no more than we should do away with coal, gas and oil electrical power plants because chemical bombs have been used to blow up buildings and planes.

Other countries have accepted the challenge of the nuclear age. France generates 75 percent of its electricity from nuclear reactors, Belgium 60, Sweden 45, Switzerland 40 and Spain and Germany each 35 percent. The Far East will be the next big users of nuclear power. Japan now obtains 30 percent of its energy from nuclear reactors. By 2010 it plans to build 40 more nuclear reactors and increase its electric power from reactors to 42 percent. South Korea has 11 operating nuclear reactors and 19 authorized or under construction, Taiwan has six operating, The People's Republic of China has four and all have plans or contracts for more. Under a new agreement, Canada will supply two more heavy water nuclear reactors to China for installation south of Shanghai.

In an editorial in the journal Science [3], Philip Abelson emphasizes the impact of the enormous expansion of nuclear reactor design, construction and operation on science in the far East. It will support a rapid expansion of nuclear scientists, engineers, technicians and manufacturing facilities while the United States faces ever decreasing nuclear energy capabilities. The number of U.S. university nuclear engineering departments has dropped from 80 in 1978 to 35 today. Unless the United States reverses its passive acceptance of misguided anti-nuclear power activists, its nuclear energy capabilities will inevitably become second class. The U.S. public has embraced nuclear medicine. There is no reason why it can't welcome increased nuclear power as well.

A Reality Check on Battery-Electric Vehicles

In November, 1996, General Motors will start selling its battery-electric EV1 through Saturn dealers in Southern California and Arizona. It will be the first electric car to reach the sales floor since the birth of the auto industry in the early 1900s. In those days battery power competed head to head with gasoline power for the future automobile market. Thomas Edison and other inventors spent years trying to perfect batteries that would compete with the gasoline engine. Primarily because of the low specific energy density, the battery-electric vehicle lost out and has never been able to compete in performance and comfort with the internal combustion engine car. It still lacks the range, acceleration, versatility and convenience demanded by the driving public.

The EV1 will do little to change the situation. It has a light aluminum frame with a modern aerodynamic design but has a range of about 80 miles and must carry lead sulfuric acid batteries weighing over 1,000 pounds. It will be priced at about $35,000. Buyers are also required to purchase a home or business electrical charger. Early buyers may expect subsidies of $5,000 from the Air Pollution Control District. They may also receive rebates from the California Energy Commission and Department of Motor Vehicles, all generous gifts from we taxpayers. The net cost is expected to be around $27,500. An alternative lease plan including charger may be available for around $450 per month.

The major argument for the switch to battery-electric cars is the lack of exhaust emissions that pollute the atmosphere. Even when the emissions by the source of electricity are included there is a claimed net gain, as electricity in Southern California is generated primarily by clean hydroelectric, nuclear and relatively clean gas.

Lester Lave, an economist at Carnegie-Mellon University and his three engineering co-authors do not accept this view. They claim [4] that substituting 500,000 electric cars for the same number of new gasoline powered ones in the year 2010 in Los Angeles would reduce the peak levels of ozone insignificantly, only 0.5 percent, from 200 parts per billion to 199. This does little to reach the current safe level of 120 parts per billion. They further maintain that in 2010 if all cars in Los Angeles were electric instead of gasoline the ozone level would be reduced only ten percent.

Some people are concerned about possible pollution of the environment during the manufacture of the lead sulfuric acid batteries--about the possible release of lead into the air, water and soil during the smelting or recycling processes. Others worry about the possible hazards of sulfuric acid spills when battery-electric vehicles are involved in serious collisions.

THERMONUCLEAR FUSION

The sun and the stars shine from the vast energies produced at their centers by thermonuclear reactions. The temperatures of ten million degrees-Kelvin are so hot that hydrogen atoms are stripped of their electrons giving a dense plasma of protons and electrons (one degree Kelvin equals 1.8 degrees Fahrenheit). Fusion reactions take place in which four protons combine to form a helium nucleus. Since the mass of the helium nucleus is slightly less than the mass of the four protons, that difference in mass has been converted into thermonuclear energy.

In other thermonuclear reactions two deuterons can react to make a helium-three nucleus and a neutron, or a tritium nucleus and a proton; and a deuteron and a tritium nucleus can react to produce a helium-four nucleus and a neutron.

Oceans contain an abundance of deuterium, almost an inexhaustible supply, so fusion supporters were optimistic that thermonuclear fusion reactors would be the preferred power source into the indefinite future. First, however, a thermonuclear reactor had to be designed and tested, and demonstrate that it generated more energy than it consumed. This turned out to be much more difficult than predicted.

In a fusion reactor it is necessary to heat the particles to temperatures high enough for them to react--temperatures comparable to or higher than at the center of the sun. In the sun the high temperature particles are confined by the sun's huge gravitational force. Since that is not available in the lab, other methods of confinement are required.

In the early days of 1953, magnetic confinement seemed near (fission reactors were developed in just two years). Various configurations of magnetic fields were used to trap the energetic charged particles and keep them away from the walls of the containers. A variety of methods for heating the deuterium or deuterium and tritium fuel to sun-like temperatures were tried. Energy break-even--to get as much or more fusion energy out of the reactor as electrical energy put in--was the goal. But the task was not easy. Particles leaked out of the magnetic bottles and instabilities developed. Tokamaks, popular magnetic confinement machines with a special configuration of magnetic fields, were developed in the Soviet Union and modified by labs in the United States and Europe. Experimenters have solved many of the heating and containment problems but commercial fusion reactors are still many tens of years in the future.

An alternative method is inertial confinement. Frozen deuterium fuel alone or mixtures of deuterium and tritium fuel in small plastic spheres in lead housings are heated by simultaneous intense pulsed laser beams from several directions. X-rays produced in the lead vaporize the plastic. This creates an implosion that increases the density and temperature of the fuel to values sufficiently high that nuclear reactions take place. The Nova laser, that can fire 10 lasers at one time with pulses shorter than a billionth of a second, was built at the Lawrence Livermore National Laboratory in 1982. But progress was slow because irregularities in the compression and other problems have been difficult to correct. This method too is tens of years from commercial use.

A one billion dollar National Ignition Facility upgrade of the Nova laser is expected to be approved by Congress. This has many more lasers beams and higher intensities than Nova that could bring the inertial confinement experiments closer to break-even. However, the primary purpose of this facility is to monitor aging nuclear weapons in the stockpile as a substitute for weapons testing. The costs are paid by a special defense programs agency within the Department of Energy.

Recent years have been difficult for thermonuclear research. The United States has withdrawn from the International Thermonuclear Experimental Reactor project (ITER) that was a cooperative venture with Europe, Russia and Japan. The largest U.S. magnetic confinement experiment, the Tokamak Fusion Test Reactor at Princeton, New Jersey will be shut down in two years. The U.S. budget for fusion research has been cut by one-third. Some of the European countries are following the United States in cutting their budgets. The ITER may also fall by the way. The future of thermonuclear fusion is now confused and in question.

Just a word about cold fusion--mentioned only because of the splash in the media and the high hopes, falsely raised, for an energy utopia. The world was thrilled by the March 23, 1989 press conference at the University of Utah where Stanley Pons, Chairman of the Chemistry Department and Martin Fleishmann, a Chemistry Research Professor at the University of Southampton, England announced they had successfully carried out fusion reactions on a table top in their lab with simple apparatus. Most surprising was the claim that fusion was successful at room temperature, and the ten million degrees-Kelvin, required by the sun and thought necessary by physicists working on fusion, was not needed.

The announcement was met with disbelief by most scientists working on thermonuclear fusion. But the prospect of unlimited energy in a small unit in a home, business or industrial plant raised the expectations of the public and many elected state and federal officials. Five million dollars were voted by the Utah State Legislature for a cold fusion institute. A hearing was conducted in Washington by the House Committee on Science, Space and Technology. Fortunately, caution prevailed as scientists from several universities and government labs convinced the committee to delay recommending federal funds for the cold fusion institute.

The Department of Energy requested their labs to determine the validity of the cold fusion claims. A panel of experts in chemistry and physics were appointed to assess the cold fusion work. Several university teams attempted to duplicate the results. Chemist Nathan Lewis and Physicist Charles Barnes lead a group of 15 scientists at the California Institute of Technology to repeat the cold fusion experiments. These checks on the Pons-Fleishmann experiments were negative. Cold fusion was dead.

Pons and Fleishmann didn't follow accepted scientific procedures. They announced their results at a press conference instead of first publishing in a peer-reviewed scientific journal. They didn't consult with their own physics department about their results or ask for advice or help. They didn't present their results at scientific meetings where the results could have been discussed and evaluated. It appeared that they were more interested in securing patents than in demonstrating that their results were correct. They forgot the advice every scientist should follow--extraordinary claims require extraordinary evidence.

1. Electrical Power

A smell of burning fills the startled Air--
The Electrician is no longer there!
Hillare Belloc, "New Digatepoem", (1910)

There is nothing like a 30-hour loss of electrical power in a March rain storm to gain respect for electricity in the home.

Lost was an oven to cook and a refrigerator to cool. The outage cut off power for the furnace, air conditioner, and even the thermostat control. The dish washer, washing machine, dryer and vacuum cleaner were useless.

Reading at night without lights was impractical. Without the radio, TV, CD player or VCR, customary home entertainment vanished. And correspondence or homework without electric typewriter, word processor or computer seemed unbearable.

We owe our comfortable home living and varied entertainment, our high standard of living and advanced level of industrialization to electrical power. And to the brilliant scientists and inventors of the 19th and 20th centuries who made it possible.

The early fundamental experiments were carried out by 19th century physicists beginning about 1820. Swede Christian Oersted, Frenchman Andre Ampere, Englishman Michael Faraday and American Joseph Henry made brilliant contributions to the understanding of magnets and electric currents and electric and magnetic fields. In 1864, the Englishman, Clerk Maxwell, merged electricity and magnetism through his famous laws of electromagnetism.

In 1878, in England, inventor St. George Lane-Fox and in the U. S. Thomas A. Edison proposed electric power for home, city and industrial lighting. Direct current--DC--systems were installed in London and New York City in 1882.

Alternating electric current--AC--systems with transformers were soon preferred because electric power could be transmitted on lines at high voltage, then decreased to lower voltage at the consumer, with considerable savings in energy.

For industrial and home use, an induction motor (to change from electrical to mechanical energy) with a rotating magnetic field, was invented in 1888 by the Croatian-born American Nikola Tesla. Westinghouse acquired the patents.

Soon after, in 1891, 100-kilowatt three-phase AC generators (that change from mechanical to electrical energy) produced electrical power in Lauffen, Germany which was then transmitted to Frankfurt. Five years later generators were installed in the United States at Niagara Falls, and power was transmitted to Buffalo. By 1898 in California, a 3-phase AC 30,000 volt, 120 km transmission line carried power from Santa Ana to Los Angeles.

Generators in the 1890s were driven by steam reciprocating engines but after 1903 by steam turbines. The steam was heated in boilers, usually by coal, to high temperatures and pressures.

After World War 1, large dams were built and hydroelectric plants constructed with generators that converted energy from water falling on turbines into electrical energy. In 1934 electrical power generated at Boulder Dam was transmitted to Los Angeles over high voltage lines at 287,000 volts.

Today, generators of electrical power use a wide variety of energy sources including the fossil fuels of coal, oil and natural gas; nuclear fission reactors; and the renewable sources hydro (dams), geothermal fluids (underground hot steam or water), solar, wind and biomass (trees and plants).

Solar can be subdivided into solar collectors that collect and concentrate energy from the sun and photovoltaics that convert the sun's energy directly into electrical energy.

In 1992 [1], coal furnished 56 percent of the electrical power, nuclear fission 22, natural gas and hydro nine each, oil three and all others combined one percent. Included in the one percent were geothermal, wind, solar and biomass. Electrical power consumed 36 percent of the total energy in the United States in 1993.

Since 1953 considerable effort has gone into development of a nuclear fusion source for electrical power. It appears that a working commercial nuclear fusion reactor will not be available before 2025 at the earliest.

2. Coal Power

The best sun we have is made of Newcastle coal.
Horace Walpole, 4th Earl of Orford, Letter to George Montagu, June 15, 1768, in Correspondence (Yale ed.) vol. 30

Many of us in Southern California are not familiar with coal.

Next to wood, coal is the oldest known fuel. Archeological evidence indicates coal was burned in Wales in funeral pyres 4000 years ago. Ancient cinders found near Northumberland in Britain show Romans burned coal before 400 AD. Hopi Indians in Southwest America heated and cooked with coal in the 12th century AD and heated their pottery kilns with coal in the 14th century.

People first mined coal from exposed outcroppings. As these were exhausted, shafts were sunk and horizontal tunnels dug. A 1684 record listed 70 mines near Bristol, England with 123 employees.

Early miners worked with pick and shovel. By the late 1800s, steam, compressed air and electricity had eased the work and reduced the dangers.

Coal is the name given to solid organic minerals, usually black, rich in carbon. Most coal in the United States was formed 225 to 350 million years ago. Large deposits of organic matter in swamps at high temperatures and pressures in the absence of oxygen led to the wide belt of coal across the eastern United States.

This coal, largely bituminous, is produced in the states of West Virginia, Kentucky, Pennsylvania, Illinois and Ohio with lesser amounts in the states of the Rocky Mountains. California is not a significant producer.

Coal is the major source of electricity in the United States and the world [1]. It produces 56 percent of the electricity in the United States; nuclear reactors, 22 percent; hydroelectric and gas each nine percent; oil, three percent and others, including biomass, geothermal, solar and wind, 0.3 percent.

Reserves of coal in the United States will last for thousands of years. When oil and gas are depleted we can turn to coal for transportation, chemicals and additional energy. So what's the problem? It's atmospheric pollution.

Coal is burned with oxygen from the air to heat the boilers. The steam produced is fed into a turbine where electricity is generated. The reaction products of coal burning are carbon dioxide and water. Since the coal is not pure carbon but contains other elements such as sulfur - additional pollutants include sulfur dioxide, nitrogen oxides and others in lesser amounts. Unrestrained, these gases go out the smoke stack.

Some of the sulfur dioxide combines with water vapor in the atmosphere to form sulfuric acid. Much diluted, it falls to earth as acid rain.

Coal power plants are responsible for about 70 percent of the sulfur dioxide, 30 percent of the nitrous oxides and 35 percent of the carbon dioxide going into the atmosphere.

The 1977 Clean Air Act Amendments mandated that flue gas desulfurization be required for new coal-fired utilities and large industrial boilers. Although costly in energy and dollars, a large number of scrubber systems were installed and working by 1988.

In response to public pressure and discussions between Prime Minister Brian Mulroney of Canada and President Ronald Reagan, the Department of Energy started a demonstration program in 1985 on Clean Coal Technologies, CCT. This provided government funds for CCT with the goal of commercialization in the 1990s.

An example of a CCT is the Cool Water Coal Gasification Project at Daggett, California. Its goal is to demonstrate the capability of producing gas from coal and generating electrical power simultaneously on a commercial scale.

In November 1990, Congress passed new amendments to the Clean Air Act. These require utilities to reduce sulfur dioxide emissions by a factor of about two and nitrogen oxides by about 20 percent by the year 2000.

Another idea for reducing pollution is to trade pollution 'allowances' like stock on the exchange. Initially the existing utilities are issued allowances on the basis of their current pollution. The utilities may use them by polluting, or decrease the pollution and sell them. Before starting up, new plants must buy required allowances.

There appears to be no way, current or projected, to prevent the emissions of carbon dioxide from entering the atmosphere.

3. PETROLEUM

Petroleum will continue to be a major source of energy far into the 21st century.

Strollers along the sandy beaches below More Mesa are surprised at the solidified tar outcroppings that oozed from the cliffs thousands of years ago.

Some are even more amazed to learn that 60 years ago hundreds of oil derricks and many short piers lined the beach and shallow offshore water from the Elwood oil field in Goleta to the Mesa field in Santa Barbara. Over 100 million barrels of petroleum (crude oil) and 90 million cubic feet of gas were produced from the Elwood field alone.

Oil has played a major role in the development of Santa Barbara County and all of Southern California. Currently California ranks fourth among states in crude oil production.

As early as 3,000 BC, the Sumerians, Assyrians and Babylonians used natural seepage of oil and asphalt at Hit on the Euphrates and other locations in construction of irrigation ditches and levees and in caulking ships and building roads.

Similar natural seepage's in the wetlands near the site of the Santa Barbara airport were exploited by the Chumash Indians.

To reduce the leaking gas into the atmosphere from the largest local off-shore seep, 1.5 miles out from Coal Oil Point, Atlantic Richfield Company built large metal tents over the leaks, then piped the gas to commercial users.

Forty percent of the total energy consumption in the United States in 1993 [1], came from crude oil, 25 percent from natural gas, 23 percent from coal, eight percent from nuclear power, four percent from hydroelectric power and 0.2 percent from all others including geothermal, wind, solar and biomass.

Transportation consumes about 40 percent of the crude oil; and residential, commercial and industrial users plus electrical power production, the rest.

Over half the crude oil used in the United States is now imported. Of that about 55 percent comes from the OPEC countries, largely Nigeria, Saudia Arabia and Venezuela, about 15 percent from Canada and Mexico each and five percent from the United Kingdom.

The U.S. 1995 Geological Survey estimated that 80 billion barrels of oil still remain in existing fields in the United States and 80 billion barrels are yet to be discovered. That oil reserve is equal to 33 years of production at the current rates. However, estimates of oil to be discovered have always been conservative.

The comparable world reserves are about 15 times higher. It is clear that U.S. and world consumption at current rates will eventually deplete the world's reserves, likely before the end of the 21st century.

Furthermore, it seems expedient to save some fraction of the reserves for manufacture of the many products derived from petroleum. During World War II the new petrochemical industry replaced many unavailable raw materials with substitutes like synthetic rubber and plastics. Today it produces fertilizers, dyes and many other chemicals for products used in the home, office, agriculture and industry.

While not unlimited, coal as a substitute for oil should last for centuries. Natural gas is an alternative, often more convenient and cleaner, that is widely used throughout the world. Its reserves, at current use, should last as long as oil.

A major problem of burning fossil fuels and biomass is pollution of the atmosphere. And the primary offender is the gasoline engine. But emissions from today's cars, per mile driven, have come down by a factor of 10 in the last 40 years. Additional reductions are needed and are expected in the future.

Emission controls on diesel trucks are necessary. Lighter cars, slower speeds, decreased accelerations and fewer miles driven will further reduce pollution. Battery electric cars are the solution only when the driver is satisfied with short range and reduced comfort and performance.

Some chemicals and alcohols added to gasoline will reduce but not eliminate pollution.

Hydrogen has been promoted as an ideal fuel. It burns oxygen to form water. It is non-polluting. However, it is not found in nature as free hydrogen but in compounds such as water. So it must first be separated from oxygen. But that takes as much energy as is reclaimed when it is burned as fuel.

4. Natural Gas

Because natural gas is the least polluting of the fossil fuels, there is the perception among many producers and financiers that it is also the least controversial. Consequently natural gas is the energy choice of the 21st century.

Natural gas production is rapidly increasing and may soon become the world's favorite fuel. It is also an important raw ingredient in the production of chemicals.

Natural gas leaks were reported as early as 6000 BC in Iran. Gas wells were drilled and gas transported through bamboo pipes in China by the year 900. Fredonia, New York had gas lights in 1821. Transport of gas by pipeline in the United States was perfected in the 1920s.

A huge increase in the use of gas followed World War II. By 1992 natural gas supplied 24 percent of the total energy in the world [1]. It was surpassed only by crude oil with 38 percent and coal with 26 percent. It is the primary energy for homes, commercial buildings and industry in the United States.

A gas plant costs about 50 percent less than a pulverized coal plant and the efficiency is greater, about 50 percent compared to 35 percent for coal. And the cost of the natural gas fuel is only about one-third the cost of petroleum fuels.

It is nearly universally accepted that natural gas is formed by thermal decomposition of organic matter under water in lakes and reservoirs at high pressures and temperatures of about 200 degrees centigrade (392 degrees Fahrenheit). It usually accompanies oil and is found as deep as 5000 meters (about 17,000 feet) on land, in offshore wells and in coal and shale deposits.

Natural gas is 90 percent methane with lesser amounts of other hydrocarbons such as ethane, propane, butane and pentane. At atmospheric pressure and temperatures above freezing only pentane is a liquid. The rest are gases but when liquefied under pressure are called liquid natural gases (LNG).

Backyard cooks will recognize liquid propane as charcoal lighter fluid and some country dwellers liquid butane as fuel for heating and refrigeration.

P. H. Abelson in an editorial in Science [5], points out that proven reserves of natural gas are about 160 trillion cubic feet (TCF) and the estimated resources about 10 times greater.

Since the annual consumption in the United States is about 20 TCF, resources used at the same rate would last for another 80 years. Past predictions have usually been too pessimistic with reserves and resources that continue to remain constant or increase from year to year.

In addition to the useful hydrocarbons in the raw natural gas, are the contaminants--water, hydrogen sulfide, carbon dioxide and nitrogen. Hydrogen sulfide is both poisonous and corrosive. It is removed with amines which may be considered organic derivatives of ammonia. Water condenses out and is absorbed by triethylene glycol.

Carbon dioxide emission into the atmosphere from methane combustion remains a problem. However, on an equal energy basis, it is only about one-half the emission from coal and two-thirds that from oil.

The gas producers, refiners and pipeline transporters, usually small companies, have banded together to support the Gas Research Institute (GRI). It has an annual budget of about $200 million for research that is distributed among universities, institutes and industrial labs. The GRI's goal is to reduce the cost of gas to consumers and benefit the gas industry with new and better technology.

Global demand for gas is expected to increase about 50 percent by the year 2010. World reserves are about 75 times the 1992 marketed yearly production. But much of the reserves are in the Middle East and the Former Soviet Union.

Large-scale international multi-company pipeline and LNG shipping contracts must be negotiated with many countries, some potentially unstable. This requires enormous capital and considerable political risk.

5. Nuclear Reactor Fission Power

If public perceptions about safety of nuclear power plants and integrity of radioactive waste repositories can be changed, and inability of elected officials to implement realistic solutions altered - nuclear reactors can furnish the world's safe, clean electrical power for the indefinite future.

The first half of the 20th century has been the golden age of nuclear physics.

In 1911, Ernest Rutherford found a small hard nucleus at the center of the atom. When J. Chadwick discovered the neutron in 1932 it was clear that the nucleus is made up of protons and neutrons. The number of protons identifies the chemical element.

Otto Hahn and F. Strasmann discovered fission in 1938. When neutrons are captured in heavy nuclei, like uranium 235 or plutonium 239, the nucleus sometimes splits into two or more smaller nuclei. This releases a large amount of energy, about a million times as much as in chemical reactions like the burning of molecules of oil, gas or coal.

It was immediately realized that controlled release of the nuclear fission energy could lead to a valuable energy source while run-away reactions could lead to a nuclear bomb. This is similar to atomic reactions where controlled burning of fossil fuels generates electric power but dynamite and TNT, in run-away chemical reactions, make chemical bombs.

Enrico Fermi, Italian Nobel Laureate in physics, foreseeing war in Europe, came to the United States in 1938. He assembled a team of scientists at the University of Chicago and on December 2, 1942 achieved controlled fission in the world's first nuclear reactor located at the Stagg Field stadium.

Why is nuclear fission power so attractive? It is because each fission of uranium 235 or plutonium 239 produces a million times the energy from each chemical fuel reaction. To produce the same amount of electrical power, roughly a million times more fossil than nuclear fuel is required.

Also for the same energy generated, fossil fuel produces about a million times more reaction products than nuclear fuel. Some of the waste products from fossil fuels such as sulfur dioxide can be removed from the smoke stack gases but carbon dioxide escapes unobstructed into the atmosphere. Since all wastes from nuclear reactors are self contained, they do not pollute the air.

The nuclear fission reactor contribution to electrical power in the United States increased from one percent in 1970 to 22 percent in 1992 [1]. It is second to coal which produces 56 percent. The United States has more reactors than any other country. However its reactor power production has leveled out over the last few years and there have been no new reactors authorized in America for 20 years.

The United States has more than 1200 operating years experience with 109 commercial nuclear power plants without a single death caused by accidents. Critics point to the Three Mile Island nuclear power plant accident, almost 20 years ago, as proof that nuclear reactors are unsafe. On the contrary, that accident is proof that nuclear power plants are safe. No one was killed or injured.

Radiation measurements of the surrounding population within a radius of five miles found an average dose of about 10 millirems. This is negligible compared to the dose of 350 millirems the average person receives in a year from background sources of cosmic rays, the ground, radon and building materials and from medical tests and x-rays. Studies to date have shown no increase in cancer rate or any other health impact caused by the accident.

More than $1 billion has been spent over 10 years to develop the Waste Isolation Pilot Project, WIPP, 2100 feet underground in the salt beds at Carlsbad, New Mexico. This is an ideal high level radioactive waste repository as the salt has been there, undisturbed, for millions of years.

Likewise, the Yucca Mountain site in the desert at the Nevada Test site near Mercury has been found suitable. The waste would be deposited in bedrock 1000 feet above the water table. A National Academy Panel determined that the water level had not risen to the repository level in the last 10 million years. At either site the spent fuel rods could be molded into glass, encased in stainless steel containers and stored safely for the desired 10,000 years.

But the Clinton Administration and Congress must act to open these sites.

Meanwhile France obtains 75 percent of its electricity from nuclear reactors, Belgium 60, Sweden 45, Switzerland 40 and Spain and Germany each 35 percent. Their reactors are mostly of U.S. design.

6. Solar Power

Solar collector power will not be a major contributor to electrical power production as far ahead as one can see. Electric power will continue to depend on coal, oil, gas, hydroelectric and nuclear fission reactor sources.

While celebrating the 25th Anniversary of Earth Day was fun and praise of renewable sources of energy satisfying, they are no substitute for a sound evaluation of the solar contribution to the generation of electrical power.

Only one percent of the electrical power generated in Southern California in 1994 came from solar power [6]. The major fraction of that power comes from solar collectors that gather light energy from the sun and reflect and focus it onto a receiver. The heat, concentrated in the receiver, produces steam that runs a turbine and generates electricity.

Harnessing the energy of the sun for man's needs is not new. We are told that Archimedes, in the 3rd century B.C., reflected sunlight from large mirrors to burn enemy ships attacking Syracuse. And Augustin Mouchot, in the late 1800s, built a solar steam engine that he gave to Napoleon III.

A 14 megawatt solar power plant was built in 1983 by Luz International Ltd. near Barstow, CA in the Mojave Desert. The company was named after the place in Israel where Jacob dreamed of building a ladder to link heaven to earth.

By 1991 it had expanded to nine plants, was generating 355 megawatts of electrical power and produced 95 percent of the World's solar energy.

Luz was assisted by a property tax credit for solar energy ventures, guaranteed sale of electricity to Southern California Edison Co. and as permitted under federal regulations produced nearly 25 percent of its electricity by the much cheaper gas-fired boilers.

Because of declining competitor oil and gas prices and less reliable federal and state tax credits, Luz International Ltd. and four subsidiaries filed for bankruptcy in November 1991. Under reorganization the plants are now operated by the Kramer Junction Company still at 355 megawatts and continue to produce 95 percent of the world market.

The other major operation in solar collector power was Solar One, located near Barstow, CA on land owned by Southern California Edison. Heliostats, nearly plane mirrors, and a receiver tower were supplied by the Department of Energy.

The cost of Solar One was about $140 million. It was a demonstration plant only, not used for power production. It operated successfully from 1982 to 1988, then was shut down but not dismantled.

Solar One has recently been modified into Solar Two where sunlight focused on the tower will heat molten salt and produce steam more efficiently than before.

A third system uses a parabolic steerable mirror with a Sterling heat engine at the focus. Seven- and 25-kilowatt systems are under construction. These too will be demonstration units only.

The major difficulty with solar power is the small amount that strikes each area on the surface of the earth. Under the best of conditions--clear skies, at the equator, sun overhead, around March 22 and September 22, it is only 1370 watts per square meter. Consequently, at a Mojave Desert site, it takes about one year's solar energy per square meter to equal the energy in one barrel of oil.

Additional problems that make solar power unattractive are the day-night, seasonal and latitude effects that reduce the power per square meter. The times of sunlight are longer in the summer and shorter in the winter. The power falls rapidly at high latitudes and when the sun is at low altitudes at early morning and late afternoon.

Furthermore, the receivers of most solar collector power systems operate at relatively low temperatures so the efficiencies for converting solar to electrical power are low, around 10 percent. The exception may be the Sterling engine with helium fluid.

7. Solar Photovoltaic Power

The major problem with generating electricity by solar photovoltaic power, at this time, is the cost of solar cells. The price must come down by a factor of 2 to 4 before solar cells can compete with fossil fuel plants and nuclear reactors, except of course, at remote sites.

Direct conversion of solar radiation into electric power by photovoltaic cells is the Holy Grail of alternative energy. There is no intermediate energy, no moving parts and it is said to be non-polluting.

Solar photovoltaic cells, commonly called solar cells, were first used on solar panels to power satellites. They are now used for power at remote sites for weather stations, telemetry systems, microwave repeater stations, freeway telephones and many other projects where power line access is difficult and cost is not an issue.

The photoelectric effect, emission of electrons by a material when illuminated by light, was discovered by Heinrich Hertz in 1887 and explained by Albert Einstein in 1905 using the quantum nature of light.

The solar cell was first demonstrated at Bell Labs in 1954. When sunlight falls on the junction between 2 dissimilar materials connected through a load, current flows in the circuit.

In 1984 ARCO Solar built the largest solar photovoltaic power plant in the world at Carrisa Plains, California. It has a maximum power output of 6.2 million watts. It was also the largest manufacturer of solar cells, shipping enough in 1988 to produce 5.5 million watts.

From 1970 to 1990 industrial companies like Chevron, EXXON, General Electric, RCA and Shell spent more than $2 billion in the solar cell industry but eventually withdrew. ARCO sold its Carrisa Plains Solar Cell Facility to Siemans. It was passed to private investors who dismantled the system and sold off the solar cell panels and guidance platforms.

ENRON and Solarex AMOCO are planning a joint five-million watt system in the Imperial Valley that would feed into the San Diego Gas and Electric grid. The Sacramento Municipal Utilities District, SMUD, is also planning two-million watt and 400-kilowatt systems.

In addition, SMUD is placing four-kilowatt panels of about 400 square feet each on the roofs of many homes in Sacramento. These feed directly into its electrical power grid. Nationwide, Photovoltaics for Utilities promotes solar cell panels on roofs of homes. These new ventures are heavily subsidized by Federal and State funds.

A large fossil fuel or nuclear power plant generates about one billion watts of electricity. Solar photovoltaic power, with less than 0.1 percent of the U.S. power production, does not compete with coal, gas, petroleum, nuclear and hydroelectric power.

Solar cell power production has many of the same problems as the solar collector method. The projected area perpendicular to the light from the sun is smaller at higher latitudes and in the early mornings and late afternoons. The sun does not shine at night and the hours of sunlight are longer in the summer than in the winter. Less light is received when the sky is cloudy or overcast.

The efficiencies of commercial solar cells are currently about 15 percent with promise of more in the future.

The average area of solar cells required to produce the same power as a large fossil fuel or nuclear power plant is then about 11 square miles.

In 1992 [1] the total electrical peak load in the United States was 695 billion watts. To furnish this power the area of solar cells would need to be 7,700 square miles. ARCO Solar at its peak was a factor of 100,000 short of furnishing the total U.S electrical power.

Since the sunlight on the solar cells varies throughout the day and night a gigantic storage system such as batteries is required.

And it's not clear that the pollution problem would be solved. With lead and sulfuric acid or other toxic chemicals in the batteries and new more efficient solar cells that might include cadmium, sulfur, zinc and tellurium, pollution could be severe.

8. Wind Power

In 1993 windpower produced about four tenths of one percent of California's electrical power. Windpower must do much better if it expects to compete in the California and world markets for production of electrical power.

The Windmill is one of the oldest machines to generate power.

It was used by the Persians in the 6th century A.D. and by the late 1100s had reached France and England. From the 12th to the 19th centuries it rapidly displaced animal power for a number of tasks such as grinding grain.

Many improvements were made with widespread use that included pumping water, sawing logs and making paper. In 1890, P. La Cour built a windmill in Denmark to produce electrical power. By the early 1900s, every farm in the U.S. Midwest had windmills to pump water for livestock and homes.

Some farms had small wind generators that charged batteries for low voltage power to radios and lights. I remember how unreliable they were. The wind wouldn't blow so the battery would run down. The lights were dim and we would miss the national radio commentary of H. V. Kaltenborn.

Liberation came when dad helped form an electrical cooperative and connected to lines of the Rural Electrification Administration.

After the oil crisis of the late 1970s and early 1980s, during Jerry Brown's administration as Governor of California, alternative energy sources of solar, wind and geothermal were encouraged. Federal and State tax breaks as high as 37 percent were given to alternative power entrepreneurs.

In addition, a market to the public utilities was mandated by the Public Utilities commission. Unfortunately, this encouraged poor design and faulty construction. When the tax advantages finally ran out, the hills of Altamont, Tehachapi and San Gorgonio passes were littered with broken-down and idle wind machines. This while the wind power fraction of electrical power in California was less than one percent.

The critics point to the monster wind machines marring the pristine desert skylines, the roar from the whirling propellers, the long distances to most consumers and the deaths of hundreds of birds of prey caught in the giant machines. Even if these difficulties can be overcome, two tough problems still remain.

The wind is erratic, even at the best sites. It can be a gentle breeze that is too still for useful power, or so strong that power production must be curtailed or stopped because of danger to propellers and turbines.

And the wind machines are made of iron and steel with moving parts. Because of the hazardous environment of sometimes strong winds, blowing sand and corrosion and often difficult and expensive replacement of equipment - the capital costs and upkeep may be too high to be competitive with other power sources.

The public is told that a new generation of wind power machines is rapidly supplanting the old, that reliable companies have replaced the overnight operations of the past and that costs of wind machines now make wind power competitive with other sources of electricity.

This is the good news. These claims must be verified. We are eagerly awaiting the evidence.

In the meantime other news is here. Good for the consumer, perhaps not so good for the wind power industry. The Federal Energy Regulatory Commission voted unanimously in February, 1995 to ask the California Public Utilities Commission to halt the auction of contracts with the independent electric power producers because it violates federal law.

Southern California Edison Co. (SCE) and San Diego Gas & Electric Co (SDG&E) claimed that the auction of contracts results in higher rates for customers. That's because the bidding process is based on 'avoided cost', a price higher than normal, that the utilities have to pay to produce or buy additional electricity.

SCE claims $1 billion and SDG&E $300 million new savings to the customer from 1997 through 2003 by the halt of the auction of contracts. The utility companies may now negotiate to purchase electricity from the independent wind power companies.

Whether this will dampen present optimism of the wind power companies is not known. In any event it levels the playing field by taking away their (along with solar and geothermal) guaranteed markets.

California's 16,000 wind machines, in 1993, produced 90 percent of all electricity in the world, generated by wind [7]. The three main sites at Altamont, Tehachapi and San Gorgonio passes provide four tenths of a percent of the total electricity used in California.

9. Geothermal Power

Geothermal energy has been tried and, at best, supplied only about three-tenths of one percent of the total U.S. electric power. It is not likely to make a significant contribution to the nation's electric power needs in the foreseeable future.

In the late 1970s and early 1980s geothermal energy, along with the other renewable energies, solar and wind, were widely promoted as the future sources of electrical power.

Geothermal energy is available at the boundaries of continental plates and other hotspots where the earth's crust is thin. Hot magma rising from the earth's interior heats the surrounding rock and water. Wells drilled hundreds and thousands of feet into prelocated sources may find hot water and steam at temperatures as high as 500 to 600 degrees Fahrenheit.

Electrical power was first produced from geothermal energy in 1904 at Larderello in Tuscany, Italy. It is still producing power today from steam originating in that natural underground reservoir. Commercial power from a hot water reservoir was produced first in New Zealand in 1958 at the Wairakei geothermal field.

Under the leadership of Governor Jerry Brown, California and the Federal Government offered economic incentives to developers of geothermal energy, who in turn promised a future of cheap, smog-free electrical power.

The largest geothermal field was established at The Geysers, 115 km (1.6 kilometers equals 1 mile) north of San Francisco. Others were opened in the Imperial Valley in Southern California and near Mammoth Lakes across the Sierras from Yosemite National Park.

By 1990 The Geysers supplied six percent of California's electric power and had 75 percent of the generating capacity of all geothermal plants in the United States [1].

However, production at The Geysers peaked in 1988 at 2000 megawatts and has been decreasing since. The steam pressure in the wells declined because the wells were running dry.

Hopes were dashed that 3000 megawatts, an amount equivalent to 3 large coal or nuclear reactor plants, could be obtained from that geothermal field.

The magma and rocks were still hot. But the water was depleted because of overdevelopment--too many wells.

In the meantime with high expectations, California had floated bonds to build two new plants near The Geysers to convert geothermal energy to electric power. The State was to construct and manage the plants, private companies would develop and operate the geothermal fields and the energy would be purchased by the Metropolitan Water District to move water from northern to southern California.

In 1981, construction was started on the $122 million Bottle Rock plant in Lake County, designed to produce electricity for at least 35 years. It was mothballed in 1990 after only five years operation. Chemicals in the steam caused pipe corrosion, lack of water reduced the pressure in the wells and the price of oil decreased so much that electricity from private utilities was cheaper than from the geothermal plants.

The second, in Sonoma County, was stopped during construction in 1985 after spending $55 million because there was not enough steam available to operate the plant.

It is even questionable whether geothermal energy should be called a renewable source. The time required for nature to resupply water to depleted geothermal wells like The Geysers is not known. It is likely to be hundreds or thousands of years.

It may be possible to slow the decline by injection of used steam or by adding water from another source. Some of this is currently being tried with mixed results.

Nor is geothermal energy smog free. The blowoff steam is often accompanied by the smell of sulfur dioxide, and excess brines with salts and minerals are sometimes stored in stand-by pools.

Other methods of generating steam such as pumping water continuously down one well onto deep hot rock and recovering hot water or steam in another has been tested with limited success at Fenton Hill near Los Alamos, New Mexico.

10. Biomass Production of Energy

Among the many problems that must be solved before biomass electrical power production can become competitive, is the intractable competition for arable land with food production, recreation and biological diversity. As yields of biomass are increased, the associated problems of water pollution by fertilizers and insecticides are intensified. It does not appear that increased biomass production will be able to solve our energy problems of the future.

Wood is people's oldest source of energy, except of course, for sunlight and the food that powers our own activities.

Fifteen percent of the world's energy is furnished by biomass--plants, trees and animal wastes--and rises to 40 percent in developing countries where much of the cooking is over open fires.

While such burning sends various polluting chemicals into the atmosphere, it is renewable since the amount of carbon dioxide absorbed in photosynthesis of growing plants is the same as that released in burning. Net reductions of carbon dioxide in the atmosphere could occur with replacement of fossil fuel by biomass. Planting biomass on deforested or degraded lands may lead to enhanced productivity, appearance and ecological diversity.

In an excellent article [8] in the 1993 edition of the Annual Review of Energy and the Environment, Eric Larson assesses biomass resources and their use in the production of electricity, fuel gas, methanol, ethanol and hydrogen.

Over 220 billion dry tons of biomass are grown yearly. However, less than 1.5 percent is used for energy. To become a significant energy source, large increases from residues, natural forests and plantations are required.

To furnish all the energy needed by the United States, about two million square miles of biomass land are required. This is higher than the land areas necessary for solar photovoltaic cells by a factor of 200 and solar collectors by a factor of 50. These large areas require considerable costs for transportation to the processing centers. In addition, biomass has lower bulk densities and less heat value per ton than coal, and high moisture content that must be removed.

Producer gas is made from biomass by thermochemical (heat and chemical reactions) and by biological (anaerobic microbe digester) processes. Biomass as an energy source was abandoned by most industrial nations after World War II, and replaced by the less expensive and more convenient natural gas.

However, producer gas is still used extensively by many developing countries. It has advantages over traditional direct biomass burning by households. It reduces indoor smoke and pollution levels that cause respiratory problems and reduces the family labor of collecting biomass.

Biogas is produced from animal and human wastes, sewage sludge and crop residues. In India and China millions of digesters serve individual homes and communities.

The biomass electric generating capacity in the United States now exceeds 8000 megawatts. This is up a factor of 10 since 1980, largely because of the incentives furnished by the Public Utilities Regulatory Policies Act of 1978 that requires utilities to purchase electricity from cogenerators. Yet, in 1992, biomass furnished less than one percent of the total electrical power generated in the United States.

For thousands of years, ethanol has been produced from plants using micro-organisms. Obtained biochemically from corn and sugarcane, and supported by government subsidies, it is now used in vehicle fuel.

Methanol and hydrogen are produced thermochemically from fiber biomass. The energy conversion efficiencies from biomass to methanol and hydrogen are about 60 percent, somewhat less than the 80 percent from natural gas.

In the United States up to 20 percent of anhydrous ethanol (zero percent water) is blended with gasoline to boost the octane rating. About 40 percent of the input corn energy is used to produce the ethanol. But to determine the net energy for conversion to ethanol, energy for producing the biomass must also be included.

In the years ahead, biogas, that drives gas turbines and later that feeds fuel cells might possibly compete economically for electric power production. But biomass fuels for internal combustion engines cannot rival petroleum fuels at present world oil prices. Someday, when coupled to fuel cell electric vehicles, the economics may be more promising.

11. Fuel Cells

Fuel cells show much promise for the future. But it is not clear at this time whether they will be competitive with other sources for the generation of electricity or with other means of powering vehicles.

Fuel cells using hydrogen gas have been praised as the energy source for the 21st century. Supporters claim it could be the most quiet, clean, efficient and cheap energy source available.

The first experiments with fuel cells date back to 1839 when Sir William Grove produced a current of electricity sufficient to shock five people holding hands.

In the 1960s, fuel cells using liquid hydrogen and oxygen with a solid polymer electrolyte (ion conductor) made up the power plant for the Gemini spacecraft. Its reaction product furnished the drinking water for the astronauts. Later spacecraft, Apollo and the Space Shuttle, used alkaline electrolytes with longer lifetimes and lower weight.

A fuel cell works much like a battery. Energy is stored in the chemicals of a battery and is released when the anode (the positive terminal) is connected through the outside circuit (car starter for example) to the cathode (the negative terminal). The main difference is that the fuel cell does not store energy, rather converts a continuous inflow of fuel directly into an electrical current.

In the simplest fuel cell, hydrogen is fed on demand to the anode and oxygen to the cathode. The two regions are separated by the electrolyte that permits the internal passage of hydrogen ions to the cathode. The released electrons form the electric current that passes through the outside circuit.

One of the big advantages of fuel cells for the generation of electricity is their high efficiency for converting chemical fuel energy directly into electrical energy.

The customary procedure with most fuels, fossil fuels for example, is to heat steam or other gas that drives a turbine to produce electricity. In this case, the second law of thermodynamics applies and limits the efficiencies for producing electrical power.

However, the fuel cells skip the heat part of the cycle going directly to electrical energy so the restrictions of thermodynamics do not apply. It is possible to obtain efficiencies about twice as high as for systems requiring heat.

The fuel cells for the space program worked well but were far too costly to compete in the commercial market. For example, expensive cryogenics were necessary to maintain the low temperatures of liquid hydrogen and oxygen.

The Energy Research Corporation has developed 100-kilowatt stacks of fuel cells for use in a two-megawatt demonstration power plant at Santa Clara, California. Each stack has 240 fuel cells with adjacent anode-cathode cell contacts. They have molten-carbonate electrolytes and are fueled by hydrogen and oxygen gases. But they suffer from a short operating lifetime and instability of the electrolyte.

Fuel cells with phosphoric acid electrolytes are available commercially. Those from ONSI Corporation at South Windsor, Connecticut have efficiencies of 40 percent and must be overhauled every five to seven years.

Cells with alkaline electrolytes have high efficiencies, and electrical current and power densities but are damaged by carbon dioxide.

Hydrogen is so reactive that it exists in nature mostly combined with oxygen as water in the oceans, or with carbon as coal, oil or gas. Therefore, hydrogen must be separated from water or hydrocarbons, then transported to the fuel cells or be produced locally by a processor ahead of the fuel cells.

Vehicles powered by fuel cells or power plants using hydrogen and oxygen would be non-polluting. The only reaction product is water. However, until hydrogen is available locally, it could be produced on board by a fuel processor to convert fossil fuels, methanol or some other hydrogen rich fuel. Oxygen would be taken from the air. In this case, carbon dioxide, nitrogen oxides and perhaps other pollutants would be present, although less abundant than from internal combustion engines.

The energy to produce the hydrogen must be considered in the net energy balance. The same amount of energy is expended in breaking up water to produce hydrogen, as is gained by reacting it again with oxygen to make water. Cheap hydrogen is required to make this method viable. That is a problem.

12. THERMONUCLEAR FUSION POWER

Thermonuclear fusion has come a long way since the buoyant days of the early 1950s. Bigger and bigger fusion machines have been built and many problems solved. But the time to a successful fusion reactor has retreated ever farther away. With the recent drastic cuts in funding, the future of electrical power from thermonuclear fusion is now in question.

What's become of thermonuclear fusion? Remember back in 1953 we were told by the optimists to expect 'break-even', more energy produced than expended, in a lab fusion reactor within a year or two.

And to look for commercial fusion reactors that produce electrical power a few years later. They would generate clean energy and because of all the deuterium in the oceans, thermonuclear fusion could be fueled for thousands of years.

Progress has been made. Scientific break-even was finally accomplished in 1994. In the lab as much energy was obtained from the fusion reactions as was fed into the plasma. But a commercial fusion reactor that puts electrical power into the grid is at least 50 years away.

Thermonuclear fusion produces energy in the center of the sun by fusing four protons into a helium-4 nucleus. When light elements are combined to make a heavier nucleus below iron in the atomic table, energy is released by fusion. About a million times as much energy is released per reaction as by chemical reactions burning fossil fuel.

For the same energy produced, the required amount of fusion fuel is down from fossil fuel by a factor of a million, as are the wastes. Also the fusion wastes are contained and do not pollute the atmosphere.

In the major fusion reaction, deuterium combines with tritium to form helium-4 and a neutron with the release of 17.6 MeV (million electron volts) of energy. The deuterium is obtained from ocean water and the tritium manufactured in the fusion reactor by the capture of neutrons in a blanket of lithium-6.

The major obstacle to the fusion of deuterium and tritium is the repulsion between the two positively charged nuclei. In order for them to approach close enough to interact, their velocities must be increased by heating to about 100 million degrees Kelvin (1 degree Kelvin equals 1.8 degree Fahrenheit).

But since all materials melt at this high temperature a material container cannot be used. Two general types of containment, magnetic and inertial have been tried.

The greatest successes to date in magnetic confinement have been the U.S. Princeton Tokamak Fusion Test Reactor and the Joint European Torus in Abingdon, England. Although reaching scientific break-even and megawatt outputs ten million times greater than the magnetic confinement experiments of 20 years ago, they are still far from the requirements for commercial reactors.

At a temperature of about 100 million degrees Kelvin, the Lawson criterion for scientific break-even of the deuterium-tritium plasma requires the product of the density of the plasma and the duration time of the burning be greater than one hundred thousand billion seconds per centimeter cubed (1 inch equals 2.54 cm).

Typical values for magnetic confinement could be a density of one hundred thousand billion particles per centimeter cubed and a time duration of a few seconds.

Future hopes for magnetic confinement depend on funding of the U.S. $500 million Tokamak Physics Experiment and the multinational $9 billion International Thermonuclear Experimental Reactor.

In inertial confinement, the inertia of the imploding ablator on the deuterium-tritium target increases its density. For scientific break-even, densities greater than about 10 million billion particles per centimeter cubed are required with confinement times one-tenth of a billionth of a second.

A tiny spherical shell capsule a few hundredths of a centimeter in diameter, containing the deuterium-tritium fuel, is compressed by a second spherical shell of high atomic number called the ablator. The ablator obtains its energy from the driver, a large number of intense laser or heavy ion beams. The ablator heats up and expands outward. This causes its inside wall to implode inward, compressing the fuel.

The Nova Laser at the Lawrence Livermore National Laboratory can deliver 10 intense beams of light symmetrically onto the tiny ablators in a time less than a billionth of a second. The Nova Upgrade at a cost of about

$400 million will carry the U.S. inertial confinement program into the 21st century.

A working inertial confinement fusion plant is far in the future. Many problems are yet to be solved. A 1000 megawatt power plant will require a repetition rate of 10 pulses per second. This puts extreme requirements on the driver, ablators and fuel targets.

13. Cold Fusion

Extreme claims require extreme proof. That was not provided. In the end, Pons and Fleishmann, and cold nuclear fusion were completely discredited.

The seventh anniversary of the March 23, 1989 press conference at the University of Utah, where two chemistry professors made the claim that electrified the world, has come and gone.

At the press conference they reported sustained nuclear fusion on a table top that produced four watts of power out, for every watt in. Stanley Pons was Chair of the Chemistry Department at the University of Utah and Martin Fleishmann a Chemistry Research Professor at the University of Southampton, England.

The electrochemists used a simple apparatus of palladium and platinum electrodes in glass cells containing conducting deuterium water solutions. They simultaneously measured neutrons from the cells. The claim had the heady promise of solving the world's energy problems.

What a convenience to have such a power source for every vehicle, home, office and business. Industry would have unlimited power. And it was clean, safe and cheap. Furthermore, it was discovered without the billions of dollars spent unsuccessfully on fusion by the United States and other major countries in the world.

By April 7, the Fusion Energy Technology Act requesting $5 million for cold fusion research was passed at the request of Governor Norman Bangerter at a special session of the Utah Legislature. The first release was $500,000 for attorneys fees to secure university patents.

On July 21, the Legislature agreed to release the remaining $4,500,000 in state funds to the new National Cold Fusion Institute.

A hearing on Cold Fusion was held on April 26 before the House Committee on Science, Space and Technology. The Utah delegation to obtain $25 million in federal funds was led by University of Utah President, Dr. Chase N. Peterson.

But caution was heard from several scientists at other universities and labs. Because of the negative reports no federal funds went to the state-supported National Cold Fusion Institute.

For the past 60 years the nuclear fusion reactions have been studied at labs around the world. It is well known that two deuterium nuclei react to give three different sets of reaction products--helium-3 and a neutron, tritium and a proton, and helium-4 and a gamma ray. The reaction rates and energies of the products have been well measured.

Knowledgeable scientists were quick to point out fallacies in the nuclear cold fusion arguments of Pons and Fleishmann. Hal Lewis, professor of physics at U.C. Santa Barbara, in an editorial in the Los Angeles Times just 10 days after the initial press conference, emphasized that nuclear fusion with excess heat at the level of four watts for a few seconds, would have given a lethal dose of neutron radiation to the two electrochemists.

Chemist Nathan Lewis and physicist Charles Barnes led a team of 15 scientists at the California Institute of Technology to repeat the electrolysis experiments. The group measured the heat output and looked for neutron, tritium, helium and gamma ray products of the fusion reactions. Their heat results were far below those of Pons and Fleishmann, and they observed no neutrons. The Cal Tech neutron upper limit was a factor of 100,000 smaller than the value claimed by Pons and Fleishmann.

A panel [9] of experts in chemistry and physics with chair, John Huizenga, Professor of Physics and Chemistry at the University of Rochester, was appointed to assess the new research area of cold fusion. The ten large national labs run by the Department of Energy were instructed to divert major resources to study all aspects of cold fusion.

The interim and final reports of the panel were published in August and November, 1989. They concluded that the cold fusion experiments did not present convincing evidence for anomalous nuclear heating or for a useful source of energy, and the evidence for the discovery of a new nuclear cold fusion process was not persuasive. Nuclear fusion at room temperature would be contrary to all understanding of nuclear reactions and would require the invention of an entirely new nuclear process.

The Panel recommended against any special funding for the investigation of phenomena attributed to cold fusion. The work of the panel and analysis of cold fusion experiments and claims can be found in the book [10] by John Huizenga, "Cold Fusion, The Scientific Fiasco of the Century," published in 1992.

In the six years since the panel final report, nothing has occurred to change the picture. Scattered groups around the world have continued their cold fusion research without convincing results. Pons resigned from the University of Utah and was reported working on cold fusion with Fleishmann in France, at the European facility of the Institute of Minoru Research Advancement, owned by an affiliate of Toyota.

Fortunately the scientific process is self corrective. The usual procedure for testing a new idea or experimental result is to try in every conceivable way to make it fail.

After discussions with colleagues, and talks at scientific meetings, it is written up and submitted for publication in a peer-reviewed journal. At each step there is feed-back, criticism and corrections of errors. Scientists anywhere may repeat the experiment or test the predictions.

Only after verification is the result accepted by the scientific community. Even then, later tests can cause it to be modified or discarded.

The secrecy of Pons and Fleishman violated all of the above checks. It appears they and the administration at the University of Utah were too intent on obtaining patents to follow the usual scientific procedures.

They first notified other scientists by means of a press conference. They published in the press. They attempted to obtain pork barrel funding instead of sending proposals to funding agencies for peer review. And they paid no attention to evaluations of review committees.

14. BATTERY-ELECTRIC VEHICLES

Because of the current high cost, it doesn't seem likely that a battery-electric vehicle will be popular as a second car for use around town. And it doesn't appear that the electric battery will replace the combustion engine in the commute, leisure or family car in the foreseeable future.

In the press it is often difficult for readers to separate the wishful thinking of dreamers and promoters from the real life results of scientific tests. This seems to be particularly true of cars and buses powered by electric batteries.

Fortunately, in Santa Barbara this is not a problem. The Santa Barbara Metropolitan Transit District (MTD) released a report in 1995, prepared by Paul Griffith, on the results of the first four years of operation of the Santa Barbara fleet of Battery-Electric Buses, the largest in the nation.

After initiation of service in 1991, the quiet, exhaust-free, odorless transportation was an immediate hit as riders increased to a million in a year.

The MTD battery-electric fleet contains ten 22-foot shuttle buses, a 22-foot transit bus and a 30-foot light-weight converted diesel bus. The shuttle buses run a downtown route over a 1.5-mile section of State Street and a waterfront route along a two-mile length of the beach. In an eight-hour day, 40 and 75 miles are covered. Typically the 22- and 30-foot transit buses service conventional MTD transit routes like Line 21, which travels 49 miles along the waterfront in 3.5 hours.

Energy consumption was found to vary up to 50 percent depending on the driver and route characteristics. It declined 70 percent when diesel propulsion was replaced by electric batteries.

A big advantage of electric battery propulsion is the absence of polluting tailpipe emissions. Even when the production of electricity at the source is considered, the pollution is only a fraction of diesel or gasoline motors.

According to a summary [10] of sources for electrical energy in Southern California in 1994, only 13 percent was produced by coal. The rest, by relatively clean natural gas--46 percent, clean nuclear reactors--21 percent and renewable biomass, geothermal, hydroelectric, solar and wind--19 percent.

Then why is battery-electric transportation not the choice of the world? The answer is the low specific energy (energy divided by mass) of batteries.

A lead acid battery has only about three thousandths of the specific energy of gasoline and diesel fuel, and three billionths of the specific energy of nuclear fuel.

The range of a useful battery-electric bus or auto is currently limited to about 100 miles because of the large weight of batteries carried along. The 22-foot bus lugs about 4,000 lbs. And the present battery charging times are long, about six hours for the MTD vehicles.

The nickel-cadmium battery, used on one bus, has a specific energy 50 percent greater than for the lead-acid battery. The performance advantage is even more because the allowable depth of discharge is about 95 percent compared to about 80 percent for lead acid. However, the capital cost per energy stored is 2.7 times higher for the nickel-cadmium battery.

Promising nickel-metal hydride and zinc-air batteries may be available for testing in the near future but others will probably not be commercially available for years.

Attempts at Battery-Electric Transportation are not new. Edison and others spent years in the early 1900s struggling to perfect high specific energy batteries. In the end the combustion engine prevailed.

The MTD has demonstrated that battery-electric vehicles can be useful where short range, poor performance and inconvenience are acceptable.

CHAPTER VI

BASIC AND APPLIED SCIENCE

The 20th century is the century of science. The discovery of the electron at the close of the 19th century paved the way for understanding the smallest objects in the universe. Research on atoms and nuclei of atoms in the first 25 years of the 20th century established the new fields of atomic and nuclear physics.

Photons of light, small bundles or quanta of energy, explained the ejection of electrons from atoms and the intensities of light at different wavelengths from hot bodies like a stove or the sun. The energy equivalent of mass, E = mc2, is a household word. Quantum mechanics was developed by E. Schrodinger, W. Heisenberg and others to describe interactions between combinations of atoms, photons and particles over small distances and short times.

The Michelson-Morley experiment showed light always travels with the same velocity in a vacuum, independent of the velocity of the source or the observer. This constant and the postulate that there is no preferred frame of reference were the bases for Albert Einstein's Special Theory of Relativity. With his theory, time dilation--an apparent longer time interval--and contraction of length--an apparent shorter distance interval--on a very fast moving object could be explained.

Einstein's General Theory of Relativity is required for motion near huge masses like the stars. It explains, for example, the bending of light that passes close to the sun. Light from a distant star appears brighter for a time while it is focused around a massive object passing between the observer and the star.

This property of the brightening of the image of a star for a time is used by astronomers in their search for dark matter--the mass in the universe that emits no light.

Scientific research at the frontier is usually not easy, straight-forward or clear. Experimenters are working at the limits of the capability of their instruments and theoretical guidance is often limited. The first experiments or observations are usually contradictory and inconsistent. It takes time to sort out the problems and carry out definitive experiments. Many different scientists contribute to the progress that often takes years or decades. Two examples are given below of research that has continued through most of the 20th century. These have contributed much to our knowledge but are far from completed endeavors.

THE ORIGIN OF THE UNIVERSE

One of the most challenging goals of basic research is to determine the origin of the universe. It has been so since the beginning of civilization. Myths have abounded as each culture, tribe and religion arrived at its own explanation. Finally in the 20th century, telescopes and instruments were designed and built with enough sensitivity to make the required observations for a scientific explanation.

Edwin Hubble, an astronomer at the Mt. Wilson Observatories in Pasadena, California found that the most distant stars and galaxies were traveling at the highest velocities away from the earth. The constant that relates the distance of a celestial object to its velocity of recession is called 'Hubble's Constant.'

In the standard model the cause of galaxies moving away from the observer is a giant explosion that occurred about 15 billion years ago--the Big Bang. Every galaxy moves away from every other galaxy. There is no center to the explosion. The galaxies move like raisins in bread dough. As the bread is baked, it rises and expands. And every raisin moves away from every other raisin.

George Gamow, Professor of Physics at the University of Colorado, and co-workers in 1948 predicted that the gamma rays produced in the first second after the Big Bang would cool as the universe expanded to a very cold few degrees Kelvin (the lowest possible temperature is zero degrees Kelvin, about -491 Fahrenheit). This radiation should permeate the universe.

Indeed, Arno Penzias and Robert Wilson, two physicists at Bell Labs, while testing radio antennas at Holmdel, New Jersey discovered this radiation coming from all directions in the sky. The background radiation left over from the Big Bang was in the microwave radio band. It had a temperature of 2.7 K, close to the value predicted by Gamow. That radiation stopped interacting with electrons about one million years after the explosion. The radiation has been traveling uninterrupted through the universe since then.

At the Planck time--the earliest meaningful time after the Big Bang, according to the standard theory, the universe was filled with all possible particles--protons, neutrons, electrons, positrons, neutrinos, quarks (the building blocks of nuclear matter) and their anti-particles. As the universe expanded, the particles cooled down and after about three minutes the abundances by mass were about 75 percent hydrogen and 25 percent helium. Smaller amounts of deuterium and other light elements also were present. Hydrogen was about 100,000 times as abundant as deuterium.

One cosmological question of great interest is whether the universe will expand forever, slow down and stop expanding or whether it will slow down and start contracting. The answer depends upon the total mass in the universe. The ratio of the mass of the universe to the mass that would just close the universe is called 'omega.' From the relative abundances of these light elements and the ratio of the number of photons to protons it is possible to determine theoretically the value of omega.

At the present time nucleosynthesis (theory of the formation of nuclei of atoms) and abundances of the light elements appear to limit omega to less than about one-tenth for normal matter. It is then speculated that the other nine-tenths would have to be exotic elementary particles left over from the Big Bang. Such particles are needed by some models of particle physics but have never been observed experimentally at accelerators or seen in observations of the stars, galaxies or other objects in the universe.

Astrophysicists using x-ray detectors on the German ROSAT spacecraft found a haze of x-rays from the gas surrounding rich clusters of galaxies. The mass of gas is many times the mass of stars in the cluster. This mass suggests an omega of about 0.2 and may be representative of the whole universe.

The largest scale used to date for observations of the universe is about 300 million light years (one light year is the distance light goes in one year). That scale is about one-fiftieth of the size of the observable universe. On this scale densities of clusters of galaxies vary significantly. Our group of galaxies is called the local group. Other concentrations are found in the Perseus-Pisces direction and in the Great Wall, the Southern Wall and the Great Attractor. The regions with concentrations are separated by regions with voids. Various computation models find omega varies from 0.15 to 1.

Cosmologists are curious about the unevenness of the radiation and matter in the early universe that lead to the formation of the galaxies. About a million years after the Big Bang, atoms were formed as electrons combined with the mostly hydrogen and helium nuclei. The photons in the background radiation were then too low in energy to break up the atoms. Any unevenness in the smoothness of the universe should show up in that radiation background and leave an imprint observable today. Astrophysicists with microwave detectors on the COBE satellite did detect a clumpiness of a few parts in a million. The galaxies formed later in the denser regions of the universe.

One model of the Big Bang explosion uses an early exceedingly fast expansion called 'inflation' by its inventor, Alan Guth at the Massachusetts Institute of Technology. The inflation model requires that the mass of the universe be equal to the mass just necessary to stop the expansion. This is the 'critical' mass. In this case omega is equal to one. Many theoretical cosmologists prefer the critical mass because of its simplicity.

One way to obtain the mass of the universe is to measure the light and other radiations from all the galaxies in the universe (actually from a few representative ones and from a few directions and on the basis of these estimate the total). Another is to observe the motions of stars in galaxies, and galaxies in groups of galaxies and from their motions and laws of gravitational attraction deduce the mass of the universe. The first gives a lower value than the second. The difference in mass between the two methods is called 'dark matter'.

The Hubble Space Telescope orbiting the earth above the atmosphere and the twin Keck 10 meter telescopes at 14,000 feet on Mauna Kea on the Island Hawaii along with the new techniques of digitizing images using charge-coupled devices--CCDs--have revolutionized astronomy. Astronomers are being inundated with mountains of data, thousands of times that available in the past. With the better resolution from the very large diameter mirrors and the better seeing above the atmosphere, images and spectra can now be measured from objects that are billions of light years away.

The tens of billions of distant new galaxies found by recent Hubble Space Telescope observations still give a luminous mass of only about one percent of the critical mass--much too low to close the universe. With gravitational lensing, astronomers are searching for dark matter in non-luminous objects--roughly the size of Jupiter called MACHOs. These are too small to burn nuclear matter and become stars. The gravitational lensing teams find that about half of the dark matter in the Milky Way halo may consist of these MACHOs. This mass is significant as the halos contain five to 10 times as much dark matter as visible. Some astrophysicists suggest that the Milky Way halo may only contain normal matter.

Our understanding of the origin of the universe is far from complete. From the number of new giant light-gathering high-resolution telescopes coming on line, the big improvements in digital imaging in position and time, and the innovational analysis programs under development to handle the huge volumes of data—in the future we can expect a vast increase in our knowledge about the origin and development of the universe.

SUPERCONDUCTIVITY

Levitated trains supported by high magnetic fields using superconductivity have been popularized as the transportation of the future. Trains would float supported only by the repulsion between their superconducting magnetic fields and the magnetic fields induced in the tracks. Tests of superconducting trains in 1979 in Japan reached record speeds of 321 miles per hour. But the liquid helium cryogenics required for the low temperature superconducting magnets was far too expensive for commercial train travel.

Superconductivity was discovered in 1911 by Kammerlingh Onnes, a Dutch physicist at the University of Leiden, who found that mercury lost its electrical resistance at temperatures below 4.2 K. A few years before he had perfected low temperature equipment that would liquefy helium. In addition to studying the properties of liquid helium, he was able to investigate the properties of other elements at low temperatures. Much to his surprise he discovered superconductivity.

Onnes' experiments were not mandated by the government. No king had ordered him to search for superconductivity. No member of the legislature required him to satisfy a societal need. The President of the University of Leiden didn't pressure him to obtain patents. Onnes was just curious.

Onnes had expectations of generating high magnetic fields with superconductivity for many applications without the usual large resistance heating and power losses. But his hopes died when relatively low magnetic fields destroyed his superconductivity. It took the next 40 years for others to develop the materials for the first successful superconducting magnets. Still there was a catch. The materials were superconductors only at very cold temperatures of a few degrees above absolute zero. It was necessary to immerse the current carrying coils in liquid helium. That was both inconvenient and very costly.

To reduce costs there was a strong motivation to find superconductors at temperatures higher than 77 K, the temperature of liquid nitrogen, so that this much less expensive coolant could be used. Slow progress was made while the critical temperatures crept upward to 23.2 K for a niobium-germanium compound in 1973. A break-through occurred in 1986 when Georg Bednorz and Alexander Muller of the IBM Zurich Research Lab in Switzerland found a critical temperature of 35 K for lanthanum-barium-copper oxide and a few months later C. W. Chu and co-workers at the University of Houston made a big jump to 91 K with yttrium-barium-copper oxide. At the time of the 10th Anniversary High Temperature Superconductor Workshop in Houston in March 1996, the record high transition temperature for superconductors was 134 K.

But the high temperature superconductor--HTC--ceramics posed new problems. It was very difficult to make wires from grains of these ceramics. When not aligned smoothly the grains prevented electrons from conducting along the wire. Recently a 50-meter underground AC transmission cable was made commercially from six kilometers of BSCCO tape--a compound of bismuth, strontium, calcium, copper and oxygen with flat, regular-shaped grains more easily aligned. Immersed in liquid nitrogen, it carried a DC current of 1,800 amps. Work on producing suitable ceramic wires is progressing rapidly at a number of industrial labs.

Applications needing low current HTS had proceeded more rapidly. Thin films of HTS used with SQUIDS--superconducting quantum interference devices--have been used for measuring extremely low magnetic fields. HTS also shows promise for filtering signals from noise in cellular phone ground stations.

The theory of low temperature superconductors was formulated by John Bardeen, Leon Cooper and Robert Schrieffer in 1957. They explained the effect in terms of the coherent motion of electron pairs that give no electrical resistance. However, this doesn't seem to explain the HTS results. Currently, there is no accepted theory of high temperature superconductivity. Progress in HTS basic research, and the development of applications has suffered appreciably because of this lack.

The Golden Age of American Science

American research scientists, at first trained in the universities and institutes in Europe then later in the United States, in the 1930s began making significant discoveries and major contributions to the world's scientific knowledge. The American universities and labs became competitive with the best in Europe.

World War II demonstrated the effectiveness of science and technology armaments and instruments of war. The best known is the Manhattan Project and the development of the nuclear fission and fusion bombs. The invention of jet propulsion and rockets gave new threats with the ability to deliver bombs first with jet planes, then V-bombs that were the forerunners of the missiles and space boosters of the Apollo moon exploration.

Radio wave astronomy has profited from the radar that was developed to detect enemy planes and to direct bombs to their targets. Private and commercial aviation is much safer now because of the many ways radio is used for navigation at night and in bad weather. Radio wave communication enabled NASA to send commands to the Pioneer and Explorer space probes as they encountered the distant planets and their moons and return data from the probes to the earth. Commercial satellites now circle the earth in stationary orbits and relay telephone messages and TV programs to customers around the world.

After World War II, the United States continued and increased its public support for basic and applied research. The science of the last half of the 20th century owes much to the famous report, "Science: The Endless Frontier," of Vannevar Bush to President Harry Truman. Bush was quite influential during the war as the Director of the Office of Scientific Research and Development.

Burton Richter, Director of the Stanford Linear Accelerator Center, Stanford, California, Nobel Laureate, and in 1994, President of the American Physical Society, in "The Role of Science in Our Society" in the journal Physics Today, quotes from the Bush report. After mentioning penicillin and radar as critical technologies with great practical benefits from long-term research, Bush said, "Advances in science when put to practical use mean more jobs, higher wages, shorter hours, more abundant crops, more leisure for recreation, for study, for learning how to live without deadening drudgery which has been the burden of the common man for ages past.....But to achieve these objectives--to secure a high level of employment, to maintain a position of world leadership--the flow of new scientific knowledge must be both continuous and substantial."

Bush then remarked, "Science, by itself, provides no panacea for individual, social and economic ills.....But without scientific progress no amount of achievement in other directions can insure our health, prosperity, and security as a nation in the modern world." Technology needs to grow along with science. Usually industry applies the science and develops the technology.

In basic research, the motivation is often to answer important questions and develop new knowledge. Little attention is given to possible practical applications. In applied research, the goal is to develop those beneficial uses. Often the research is not clearly one or the other but changes with progress and time.

Richter points out, "Science enables industry to reduce scientific discovery to practical applications effectively and quickly......There must be a continual interaction between scientists in the laboratory and engineers in industry." He points out that the modern laser technology had to await the theoretical basic quantum mechanics of the 1920s-1940s including the

Einstein stimulated emission and absorption coefficients. The transistors of the 1950s followed the basic condensed matter research of the 1920s and 1930s, and the medical magnetic resonance imaging of the 1980s was based on the nuclear magnetic studies of the 1930s and 1940s.

The basic policy of the United States has been to fund basic research primarily in universities, to fund government labs and encourage private industry through the National Science Foundation in the physical sciences, the National Institutes of Health for the biological and medical sciences, NASA for the space sciences and the various federal departments like agriculture, commerce, defense, energy and transportation for other specific disciplines.

The key to the excellence of U.S. research has been the competitive review of proposals submitted for grants. Only the best proposals, as evaluated by external review individuals and committees of peers, are funded. This method is a departure from the traditional European funding of institutes as a whole.

Universities in the United States are now considered by many educators among the best in the world. Our graduate programs attract students from all over the earth. In fact, half of the graduate students attending U.S. graduate school in science disciplines are from foreign nations. Many remain in the United States and contribute to high quality research in our universities, government labs and in private industry.

There are various ways to measure the success of basic research. Each has its faults and biases. One way is by the awards of Nobel Prizes in the fields of physics, chemistry and physiology-medicine--the most prestigious awards in science. Of the total 473 Nobel Laureates since the first in 1901, 42 percent have been U.S. scientists. In the early years from 1901-1930 and 1931-1945 only 7 and 29 percent were American. During that time the United Kingdom, Germany and France were the world leaders in awards as they were in basic research and quality of graduate student education. For the next 30 years about half of the awards went to Americans and in the last 20 years the fraction has climbed to about two-thirds. In 1995 seven of the nine awards went to the United States.

It takes time for new discoveries to be verified and new areas of research to be recognized, so the Nobel Awards are often given long after the time of the noteworthy research. The lag in the Nobel Awards varies but may be as much as 20 years on the average. There are some critics who suggest that U.S. science has reached its zenith, and has already started to decline.

IS AMERICAN SCIENCE IN DECLINE?

In the journal Science, P. H. Abelson discusses "The Changing Frontiers of Science and Technology" [1]. He raises questions about the future of U.S. science brought on by global industrial competition. Major concerns are that the United States now has the world's greatest international debt and the largest trade deficit. It continues to lose supremacy to East Asian countries in a number of critical areas. These world problems are affecting American universities, federal labs and private industry.

During the 1970s, Abelson points out, the United States maintained a big trade surplus by exporting agricultural products, automobiles, petroleum and steel. Major corporations like Bell Telephone, Dow, Dupont, General Electric, Hewlett-Packard and IBM generously supported their own labs.

On visiting 20 of these labs Abelson found, "Research and development were leading to new products, and long-term studies designed to understand nature were still fashionable. The level of scientific competence was excellent, and to achieve rapid progress toward goals, the best companies assembled interdisciplinary teams of scientists, engineers, and marketing experts. The morale of team members was excellent, and staffs faced the future with confidence."

Unfortunately these pleasant days have disappeared. Like many other industrial organizations, these corporation labs have been down-sized or eliminated. According to Abelson, support of long-term research dropped from 6 percent of R&D expenditures in 1988 to 1.8 percent in 1994.

According to the report, "Critical Technologies Update 1994" of the Council on Competitiveness, of 94 critical technologies the United States was strong in just 31 and competitive in 42. Strengths were in genetic engineering, computers, computer-aided engineering and information technologies.

The United States has lost its lead in the energy field. In the middle of this century it was dominant in petroleum exploration, production, refining and in the manufacture of petrochemicals. It now imports 50 percent of its oil and refined products.

The same is true for nuclear energy. No contract has been let for a commercial nuclear reactor in the United States in the last 20 years. The initiative in nuclear energy has moved to France, Japan, South Korea and other countries in East Asia. Large numbers of science and engineering students are being trained in the new universities there. Abelson reports that in 1990, six Asian countries produced more than 250,000 first-degree engineers--compared with 65,000 in the United States.

In the Far East many countries are expanding their R&D with the widespread conviction that supporting science and technology is the highway to a better future.

Universities are facing new realities. Federal funds for research grow ever tighter. Research suffers as staff support is down-sized. Ever increasing societal demands on the states result in smaller fractions of the state budgets going for graduate student education.

A more optimistic tone is voiced by Eugene Wong, Hong Kong University of Science and Technology, Kowloon [2]. He states "Contrary to the anecdotal evidence of the 1980s, basic research does confer a preferential economic advantage on countries that fund it. That is why the United States is likely to dominate vital markets well into the next century."

It is commonly argued that Japan is a country that has benefited from basic research carried out by the United States, Wong says. "Japan has emphasized product development and high-quality manufacturing. In so doing, it had captured, by the late 1980s, one after another of the science-based markets: television sets, video recorders, liquid-crystal displays and dynamic random access memories (DRAM)--all technologies invented and commercialized by the United States."

It is further said that other countries on the Pacific Rim, Taiwan, Hong Kong, Singapore and South Korea are rapidly growing, using strategies similar to Japan. This model runs counter to the Vannevar Bush basic research model used successfully in the United States which is now under attack and may be in jeopardy.

Instead Wong points out, "Japan, with twice the capital investment rate (of the United States), should have more than twice the rate of economic growth. But that is not the case. The high rate of growth in Japan in earlier years has slowed dramatically. Even after it recovers from its current recession, the sustainable rate of growth in Japan is estimated to be no more than 3 per cent of Gross Domestic Product a year, whereas the comparable estimate for the United States is 2.5 per cent. So despite its greater R&D investment, Japan's economy is considerably less, not more, efficient."

Wong continues, "Universities and basic research in the United States have been major beneficiaries of its post-war science policy. By any reasonable measure, they have reached heights never previously scaled. Their superiority over their counterparts in Japan cannot be questioned......this superiority has contributed in important ways to the efficiency of the U.S. economy."

Switzerland is a small country that has continued to thrive with a healthy mix of basic science and manufacturing notwithstanding its high wages. It has succeeded largely because of its renowned universities and first class basic research.

In conclusion Wong writes, "The U.S. semiconductor industry did not waste away, and while policy pundits were mourning the loss of the consumer electronics market, U.S. researchers were creating revolutionary markets as diverse as biotechnology, multi-media, computer software and digital communications. None of these would have existed without federally supported research, and none would have existed without the direct participation of some of those who did the research. U.S. industry is now in a position to dominate these vital areas of the economy well into the next century."

To paraphrase Wong: it's been said to foretell the future, one has to invent it--to invent it requires basic research.

1. The Big Bang Origin of the Universe

We are fortunate to live in one of the most exciting times in the history of astronomy and in the study of our universe.

The question of the origin of the universe is one of the most challenging in astronomy and perhaps of all knowledge. It was asked by the ancients, by the shepherds who looked up at the stars at night and by the scholars through the centuries until now.

A major step forward occurred with the invention and use of the optical telescope by Galileo 400 years ago. With larger telescopes that could gather more light and resolve dimmer stars, 20th century astronomers made rapid progress.

Edwin Hubble showed in the 1920s that many 'nebulae' were actually galaxies, made up of billions of stars, much like our sun. He discovered the first evidence for the expansion of the universe.

As the universe expands every galaxy gets farther away from every other galaxy, and the more distant ones move away faster than the closer ones. The velocity of expansion is equal to 'Hubble's constant' times the distance away. The inverse of Hubble's constant gives the age of the universe, about 15 billion years.

Since space is expanding, photons of light coming toward us have their wavelengths stretched or lengthened. We see longer wavelengths and say they are red shifted. The photon's energies are reduced.

In the 'Standard Model' of the origin of the universe, a huge explosion--the 'Big Bang'--occurred about 15 billion years ago. Prior to that, space and time were all tangled up. The laws of physics were not valid. However, an infinitesimal time after the Big Bang--at the 'Planck time'--the laws of physics became valid and the universe has been expanding ever since.

In 1965 Arno Penzias and Robert Wilson at Bell Telephone Labs were testing a large horn-shaped antenna at Holmdel, New Jersey to reduce the background noise to the lowest possible level. They pointed it at the sky to reduce the noise of moving cars, TV and other man-made interference.

No matter in what direction the antenna was aimed, they were unable to reduce the noise below about 3 degrees above absolute zero. Perplexed, they discussed the result with their colleagues.

Interestingly, only a few miles away at Princeton, Robert Dicke and associates had predicted there should be a cosmic microwave background radiation left over from the Big Bang and were designing equipment to look for it. Actually George Gamow and co-workers back in 1948 made similar predictions, but they were apparently lost in the literature.

At the Planck time, just after the Big Bang, the radiation in the universe was extremely hot, at a temperature of about 3, followed by 32 zeros, degrees.

As the universe expanded the temperature cooled so that today it is only the 3 degrees Kelvin (K) observed by Penzias and Wilson.

Recently, the cosmic background explorer satellite COBE, with very accurate measurements, found the temperature to be 2.726 K. The background radiation deviated from isotropy by only a few parts in a million. This deviation is enough to allow the formation of the galaxies.

The mass abundances of the elements in the universe have been accurately determined in several ways to be about 75 percent hydrogen, 25 percent helium and 1 percent all other elements. Hydrogen is about 100,000 times as abundant as deuterium. The photon to proton ratio now is about one billion. These measured abundances are consistent with the Standard Model.

At the Planck time the universe was filled with all possible particles, protons, neutrons, electrons, positrons, neutrinos, quarks (the building blocks of nuclear matter) and their anti-particles. Initially the four forces--strong, electromagnetic, weak and gravitational--were all equally strong. But as the universe expanded each of the forces changed to the strengths measured today.

One of the questions for the future of the universe is whether it will expand forever, stop expanding and keep its size, or stop and start contracting. This depends on the amount of mass in the universe.

Current observations of the visible mass suggest that the mass of the universe is only a few percent of that necessary to stop the expansion. However, measurements of the motions of stars, galaxies and clusters of galaxies imply that there is more mass in the universe than measured with telescopes. Astronomers and astrophysicists in many countries are actively searching for the additional dark matter.

We can expect many new measurements and observations in the next few years from the Hubble telescope and new larger ground telescopes that may answer this question. The Standard Model will then receive additional confirmation or important modifications.

2. The End of the World

Life on earth does not appear to be at risk from natural causes for tens of millions of years. Whether life on earth can survive the impact of people for this time is another question.

According to a recent U.S. News and World Report [3], "Nearly 6 in 10 Americans believe the world will come to an end or be destroyed, and a third of those think it will happen within a few years or decades."

Before packing their bags these people might want to consider some known possibilities for the "end of the world."

We can quickly eliminate collisions of other stars with the sun as there have been no collisions in the last 5 billion years and calculations show none are expected in the next 15 billion years, the age of the universe.

In about 5 billion years, after it has consumed its hydrogen fuel by fusion, the sun is expected to expand into a red giant star. It will engulf the earth, kill all life and burn the earth to a cinder. Perhaps in a billion years or so, the sun's radiation will increase enough and the earth's temperature rise sufficiently that life will be unpleasant if not impossible.

No immediate worry here.

Likewise no problem with other planets or the moon. The planets will remain in orbits around the sun and will not collide with the earth. The moon will continue orbiting the earth, gradually slowing down and moving away.

However, collisions of comets and asteroids with the earth have occurred in the past and will continue in the future.

Thousands of meteors, pea size, enter the earth's atmosphere every year and are seen as 'shooting stars'. They are heated by friction with the earth's atmosphere and most burn up before reaching the ground. Larger meteors land on the ground, or snow and are later picked up as meteorite finds. The largest ever is the 60 ton Hoba West, found near Grootfontein, Namibia.

Two large asteroids have collided with the earth in the last 100 years. On June 30, 1908, near the Tunguska River in Siberia, an explosion flattened more than 1,000 square kilometers of forest and killed many reindeer. No impact craters or meteorite fragments were found.

Most likely a 100,000 ton stony meteorite exploded in the atmosphere with the blast of a 10 megaton nuclear bomb, and the shock waves caused the damage.

On February 12, 1947, near Vladivostok, Siberia, fragments of an asteroid with a fireball as bright as the sun struck the ground and dug 106 craters up to 28 meters across. More than 23 tons of iron meteorite fragments were recovered. These two asteroids could cause considerable damage if they crashed near populated areas. However, asteroids of this size are no threat to end life on the earth.

Remember the excitement in July, 1994 when pieces of comet Shoemaker-Levy 9 crashed into Jupiter. Bright spots persisted for hours. Some of the 21 fragments may have had diameters as large as 2 kilometers. The plumes of debris splashed thousands of kilometers above Jupiter's surface.

Jupiter is much more massive than the earth so the chance of a comet like Shoemaker-Levy 9 striking the earth is much less than for Jupiter.

However, the earth has been hit by comets and asteroids in the past. In its initial formation, the earth suffered intense bombardment from debris in the solar system. But that was over 3.5 billion years ago.

The youngest large crater on the land surface of the earth is the Meteor Crater in Arizona, just off U.S. 40 between Flagstaff and Winslow. It is 50,000 years old, caused by an asteroid about 100 meters across.

In 1980, geologist Walter Alvarez; physicist Nobel Prize winner Louis Alvarez, his father; and their coworkers startled the world by suggesting that the collision of a giant asteroid with the earth was responsible for the death of the dinosaurs, 65 million years ago.

The evidence was a layer of iridium enriched sediment of the proper geological age found at many sites on the earth. Recently, the impact site was located near Chicxulub, in the Mexican Yucatan Peninsula.

The diameter of the asteroid is estimated to be about 20 kilometers. Its mass was more than one trillion tons and the energy release equivalent to 100 billion megaton bombs.

The cause of the dinosaurs' demise was a cloud of sulfur particles blasted into the atmosphere that formed sulfuric acid clouds. The umbrella of clouds over the earth blocked out the sun for many years, long enough for the resulting freezing climate to destroy the dinosaur food supply.

The time to the next such catastrophe is unknown but should be measured in tens of millions of years. By that time we should have developed a strategy for defending against such a collision.

For example, such an object could be detected and its trajectory tracked accurately for years. With technology, little advanced from today, it should be possible to divert its path away from the Earth with nuclear explosions at safe distances of hundreds of millions of kilometers from the earth.

3. THE STRUCTURE OF MATTER

The quark was named by Nobel Laureate Murray Gell-Mann, from "Three quarks for Muster Mark!" in "Finnegans Wake" by James Joyce, the Irish novelist.

On March 2, 1995, the Department of Energy Fermi Lab at Batavia, Illinois announced the discovery of the top quark. While that news, for most readers, didn't compete with the House Republicans' first 100 days, for elementary particle physicists it completed a search for the Holy Grail.

Much earlier in the 5th century B.C., the Greek philosophers speculated that all matter was composed of a few simple indivisible particles arranged in different combinations and amounts.

In the early 1800s, the English chemist, John Dalton, demonstrated that atoms could combine in different ways and proportions to form molecules. For example, nitrogen (N) and oxygen (O) could combine to form nitrous oxide (two Ns and one O), nitric oxide (one N and one O), nitrous anhydride (two Ns and three Os) and nitrogen oxide (one N and two Os).

Ernest Rutherford, founder of nuclear physics, and his co-workers, in 1911, scattered alpha particles from gold and other atoms. They discovered that the protons were concentrated in a small region near the center of the atom, called the nucleus. Electrons traveled in orbits around the nucleus like a tiny solar system. Most of the space of the atom was empty.

The description of the atom was completed in 1932 when the English physicist Nobel Laureate James Chadwick discovered the neutron.

The simplest atom, hydrogen, has one proton in the nucleus and one electron in the ground state (circular orbit around the nucleus). Deuterium, an isotope of hydrogen, has one proton and one neutron in the nucleus and one electron. Helium 4 has 2 protons and 2 neutrons in the nucleus and 2 electrons. Uranium 238 has 92 protons and 146 neutrons in the nucleus and 92 electrons.

The diameters of atoms and nuclei are very small, the hydrogen atom about 0.00000001 cm (2.54 cm equal 1 inch) and the nucleus (proton) about ten millionths of that. Other atoms and molecules are slightly larger.

With the cyclotron and other accelerators, new unstable particles including pions, kaons, charm mesons, the Psi and the Upsilon were discovered. Two new forces, the nuclear or strong force and the weak force were added to the already known electromagnetic and gravitational forces.

In 1964 Gell-Mann and G. Zweig proposed that neutrons, protons and mesons, themselves, were made up of smaller more fundamental quarks; protons and neutrons of 3 quarks and mesons of a quark and an antiquark. Charged antiparticles are identical to particles except for opposite charge, i.e., positive instead of negative, or vice versa.

The standard model of elementary particles has been confirmed. It includes: the 6 quarks-- up, down, strange, charm, bottom and top that connect with all 4 forces and the leptons-- electron, muon, tau, electron-neutrino, muon-neutrino and tau-neutrino, that connect to the weak, electromagnetic and gravitational forces.

Also included are the photon that carries the electromagnetic force, W-plus, W-minus and Z-zero that carry the weak force, the gluon that carries the strong force and the graviton (postulated not yet discovered) that carries the gravitational force.

An additional particle, the neutral Higgs boson, is predicted by the standard model but has not been found. Higher energy accelerators are required to produce its mass. Elementary particle physicists hope, in a few years after the Large Hadron Collider at CERN near Geneva, Switzerland is modified, the Higgs particle will be detected.

4. ANTIMATTER ATOMS

Apparently we live in a normal matter universe. Early in time, in a fraction of a second after the Big Bang that occurred about 15 billion years ago, most of the matter and antimatter annihilated. And the slight excess of matter over antimatter that existed then, is the universe we live in today.

The formation of the first antimatter atom was announced in January, 1996. Eleven antihydrogen atoms were produced at the European Laboratory for Particle Physics (CERN) in Geneva, Switzerland.

An antihydrogen atom is formed from an antiproton and an antielectron, now called positron. The ordinary matter equivalent is the common hydrogen atom made up of a proton and an electron.

The positron was predicted by P.A.M. Dirac, an English theoretical physicist in 1930. He conjectured it should have the same mass as an electron and the same charge, but positive instead of negative. For this, he received the Nobel Prize in physics in 1933.

In 1932 the postulated positron was discovered by Carl Anderson at the California Institute of Technology, Pasadena, in cloud chamber pictures of cosmic rays. A magnetic field in his cloud chamber detector caused positrons to bend in the same direction as the protons, as both are positively charged. Electrons with negative charge bent in the opposite direction. The Nobel Prize in physics to Anderson followed in 1936.

Every particle has an antiparticle. Particles and their antiparticles are formed in pairs. They are also destroyed in pairs. Gamma rays with sufficient energy can collide to form an electron-positron pair. Likewise when a positron finds an electron they are annihilated and their mass energy is changed back into gamma rays.

The Bevatron accelerator at the Radiation Lab at UC Berkeley, completed in 1955, was designed to accelerate protons to energies of 6 billion electron volts--enough energy to produce proton-antiproton pairs. In the first experiment carried out at the Bevatron, antiprotons were discovered by Emilio Segre and Owen Chamberlain. They were awarded Nobel Prizes in physics in 1959. But the antiproton had no positron attached to make it an antihydrogen atom.

Antiprotons as well as positrons have been detected in cosmic rays. Positrons are fewer than electrons by a factor of ten and antiprotons are down from protons by a factor of 2000. Both positrons and antiprotons are made by interactions of energetic protons and other particles with matter in our galaxy.

To assemble the antihydrogen atoms, a team led by German physicist Walter Oelert, from the Institute for Nuclear Physics in Julich, Germany, used the Low-Energy Antiproton Ring at CERN. Antiprotons, produced by high energy collisions, are injected into the ring where they coast for many revolutions.

Electron-positron pairs are generated by occasional scatters of the antiprotons in the Ring. When the velocity and position of a positron and antiproton are sufficiently close, they combine into an antihydrogen atom. The antiatom is neutral and not bent by the magnetic field so flies straight ahead and is detected by its annihilation with normal matter. In the process its mass energy changes into gamma rays and mesons that are counted in particle detectors. The antihydrogen atoms last only 40 billionths of a second.

To construct other kinds of antiatoms is a formidable task. Even the simple deuterium isotope of hydrogen requires the same velocities and positions for an antiproton, antineutron and positron. And the antihelium atom requires the same for two antiprotons, two antineutrons and two positrons.

Unfortunately, antimatter does not appear to be a useful energy source. Neglecting the inevitable energy losses that occur in the generation of energy, the same energy is expended to create the antimatter as is gained in its annihilation. Of course, no lumps of antimatter could exist on or in the earth. They would have been destroyed by interactions with normal matter as the earth was formed.

Furthermore, no antimatter asteroids, comets or planets in the vicinity of the solar system have been observed that could be tapped for energy by annihilation with normal matter. In fact, such objects would have been destroyed long ago by bombardments of normal matter.

There is small chance that isolated pockets of antimatter could be preserved anywhere in our galaxy—nor is it likely that antigalaxies exist in our universe.

5. Molecule of the Year

The new state of matter opens a wide range of investigations. In the coming years we may expect many new discoveries and applications as the techniques of Bose-Einstein condensates are applied to a majority of elements in the atomic table.

The journal Science each year names a "Molecule of the Year" [4]. In 1995 the honor went to the recently discovered state of matter, the Bose-Einstein condensate.

The condensate, discovered by Eric Cornell at the National Institute of Standards and Technology and Carl Wieman at the University of Colorado, both at Boulder, Colorado and their colleagues confirms a prediction made 70 years ago—that large molecules of integer spin (angular momentum) particles should exist.

The prediction was a consequence of the statistics of particles developed by the Indian mathematician and physicist Satyendra Bose and the well-known Albert Einstein. It applies to 'Bose-Einstein particles', those with integer spin. In the theory of particles any number with integer spin may occupy the same quantum mechanical state.

The key to the success of the Boulder physicists was the invention of a refrigerator to cool rubidium gas atoms to the coldest temperature ever recorded, 170 nanokelvin--170 billionths of a degree above absolute zero.

At this cold temperature, many of the atoms are in their lowest possible energy states. According to quantum mechanics these atoms overlap. They act together in union, like a large molecule.

The race for the discovery of the condensate was won by the Boulder group [5] when they combined several cooling methods to reach the required low temperature of the atoms in the gas.

The authors reduced the velocities of the atoms with a rubidium laser beam. The laser itself is an example of quantum mechanics applied to light. The rubidium atoms of the laser, locked together in phase (the emitted light waves vibrate together) emit light at a single frequency. By shifting the frequency slightly downward, the approaching low temperature atoms see the light, Doppler shifted upward to the resonance interaction frequency. When low temperature rubidium atoms absorb the light photons they are slowed down and cooled to an even lower temperature.

The Doppler shift is the change in frequency at the observer when the source of light or sound approaches or recedes. The pitch of an approaching police siren goes up, and of a receding one, goes down.

Although the laser cooling reduces the temperature to about one millionth of a degree above absolute zero, further cooling by evaporation was found necessary. The laser cooled atoms were transferred to a magnetic trap in a magnet that has a strong magnetic field at the edges but weak in the center. The hottest, fastest atoms were allowed to escape and the slower, cooler atoms were left behind. It is much like the cooling of a wet towel following drying-off after a shower.

Still one problem remained, since the magnetic field was zero at the center, the atoms didn't stay trapped but continually leaked out. As a final solution they added a second magnetic field that swung the zero magnetic field point around in a circle. This second trap then acted like a spinning "top", revolving around the geometric center. The atoms inside the circle of the revolving zero point saw a magnetic field so were trapped and continued to cool down.

The resulting condensate consisted of a ball about 10 microns in diameter containing a few thousand atoms (one micron is 4 hundred-thousandths of an inch). When photographed with laser light the physicists saw only a smear. By reducing the trapping to zero and permitting the gas condensate to expand for about 60 milliseconds, the cloud became sufficiently transparent for laser photography.

The velocity distribution of the atoms was measured and a near zero velocity Bose-Einstein condensate spike was detected at the center surrounded by a diffuse non-condensate cloud of atoms. A condensate of 2000 atoms could be preserved for 15 seconds. Much larger condensates lasting for longer times with other atoms have more recently been produced by the Boulder and other groups.

In addition to the condensates, the Bose-Einstein statistics also applies to zero spin particles of liquid helium that climb unrestricted over the walls of beakers. And to superconductivity of currents in selected wires at low temperatures explained by electron pairs that act as particles with integer spin.

6. The Earth's Atmosphere

While driving up the mountains this summer, to avoid a blowout, you may have found it necessary to let air out of your tires. This was needed to keep the difference in pressure between the inside and outside of the tires constant. After descending, you pumped up the tires again.

The air--or atmosphere--of the earth extends upward for hundreds of miles. Its pressure decreases by about a factor of 2 for every 10 miles increase in altitude.

At heights of several thousand feet you may have become tired, listless and irritable. This is caused by the reduced atmospheric pressure giving less oxygen in each breath of air. At the peak of Mt. Everest, 29,028 feet high, it is extremely difficult to function. The atmospheric pressure is only one-third that at sea level.

Because water boils at lower temperatures at the lower pressures at high altitudes, it takes longer than at sea level to cook a hard boiled egg.

The earth's atmosphere is about 0.78 nitrogen, 0.21 oxygen and 0.01 argon, water vapor and carbon dioxide. It was formed about 2 billion years ago from outgassing of compounds accreted onto the earth by collisions with rocky asteroids and icy comets.

When air is breathed by animals, the oxygen reacts with the hydrogen and carbon in the food they eat. Energy is released in the chemical reactions that form carbon dioxide and water. These reactions in the body are much like those occurring in burning fossil fuels or wood.

Much of the released carbon dioxide is consumed by plants in photosynthesis with energy furnished by light from the sun. In this process oxygen goes back into the atmosphere.

Lots of carbon dioxide has gone into the shells of sea creatures. The shells settled to the bottoms of the oceans where beds of limestone were formed. As plants were compressed at high pressure in the bottoms of lakes, in the absence of oxygen, oil and gas deposits were created.

After a shower the bathroom mirror steams up because it is covered by tiny droplets of water. Hot air can hold more water vapor than cool air. When the hot air from the shower, saturated with water vapor, comes in contact with the cooler mirror, the air cools and water condenses out onto the mirror.

Coastal fog is caused by the upwelling of cold seawater from lower depths that comes in contact with warmer air above the water. Water droplets condense out as fog.

We are all familiar with evaporative cooling. When hot, you perspire and your skin cools. It takes heat or energy to evaporate water, to change it from a liquid to a vapor. That energy or heat is taken from the skin by the evaporation process.

The evaporated water from oceans, lakes and moist soil rises to form clouds. Continental and intercontinental jet planes usually fly at altitudes higher than 35,000 feet to avoid the turbulent weather in the clouds below. A large amount of energy as well as water is stored in the clouds.

A bolt of lightning occurs when the air is ionized by a short burst of electrons that flows between clouds at different electric potentials. The thunder is the sound generated by the lightning. Sound in air travels at about 350 meters per second--780 miles per hour.

Since the light arrives almost instantaneously, you can determine the distance to the lightning flash by measuring the time interval between the arrival of the light and the sound. A time of 4.5 seconds indicates the lightning flashed about 1 mile away.

The sudden release of enormous amounts of stored energy in the clouds can sometimes cause large storms and hurricanes. The energy is released when the water vapor in the clouds condenses into rain. As a hurricane passes over warm tropical water it can pick up additional water vapor that feeds the clouds and increases the strength of the storm.

7. Earthquakes and Continental Drift

The 6.7 magnitude Northridge earthquake on January 17, 1994 and the 7.2 Kobe, Japan earthquake one year later have sensitized the public to earthquake dangers and cost to life and property.

Recent reports in the Los Angeles Times and Santa Barbara News Press have identified risks of earthquakes to Santa Barbara and other communities in Southern California.

Although it doesn't neutralize the risk, it may reduce the trauma to understand why California is in this fix. Just blame it on the Continental Drift.

Young students in geography, studying maps of the earth for the first time, are quick to point out that the continents can be moved to fit together like a jigsaw puzzle. The bump of Brazil in South America fits neatly into the shoreline of middle Africa. The North African bulge nestles against the Atlantic coast of North America, and Greenland fills in the northern gap between North America and Europe.

In 1912, the German meteorologist Alfred L. Wegener, suggested that the continents once formed a single land mass that he called Pangaea, and have since drifted slowly to their present positions. This idea was not accepted for nearly 50 years because it was thought that the crust of the earth was rigidly attached to the mantle and such movement was impossible.

Evidence from matching ancient plant and animal species on the separated continents strengthened Wegener's theory. The definitive evidence came from radioactive dating of sea floor materials near the mid Atlantic Ridge. The North and South American plates were found receding from the Eurasian and African plates at speeds of a few centimeters per year (2.5 centimeters equal 1 inch).

In addition, hot lava containing iron squeezed up from the interior of the earth between the plates, cooled and solidified. It left a record of the direction of the earth's magnetic field that switches from north to south and back over times of thousands of years.

With the development of the new theory of plate tectonics in the 1960s, geologists embraced the continental drift.

The solid crust and upper mantle of the earth, called the lithosphere, divides into several continental plates. These plates float on the pliable upper layer of the mantle called the asthenosphere.

Hot lava oozes up at the ridges and pushes the plates apart. The opposite side of the continental plate jams against another plate approaching from the opposite direction.

The lower denser oceanic plate burrows under the continental plate and crust material is returned to the interior of the earth. Such subduction is found at the Japan trench.

The upper plate rises to form mountains. The Himalayas, the highest mountains on earth, are still being formed by the collision of the Indian and Eurasian continental plates.

And one plate may slide parallel to another, as the North American plate against the Pacific plate to form the San Andreas fault. In a few million years, Los Angeles riding on the Pacific plate may end up a coastal island off San Francisco.

The boundaries between the plates are the sites of intensive earthquakes and volcanoes. As one plate tries to slip against another, friction causes intense stresses to build up that are suddenly released in terrifying earthquakes.

Volcanoes are formed when hot magma in the mantle breaks through the crust as lava, flows out then cools to form cinder cones and basalt plains. These arise at mid-ocean ridges, subduction trenches and mountain ridges. They are also formed above mantle hotspots. The Hawaiian island chain was formed during the last 100 million years as the Pacific plate passed slowly over the Hawaiian mantle hot spot.

The time since the break up of the supercontinent Pangaea, is estimated to be about 300 million years. The motion of the continental plates away from Pangaea cannot continue indefinitely. Eventually they bump against neighboring continental plates.

It is likely that previous cycles of continental drift occurred. Old eroded mountain belts like the Appalachian Mountains are likely relics of a previous cycle. Each cycle may last up to 600 million years.

The Atlantic spreading may now be near its end. It is speculated that the North American and Eurasian plates may reverse their directions sometime in the next 100 million years.

Then Californians may obtain relief. The volcanoes and earthquakes may shift from the Pacific rim to the Atlantic coasts of North America and Europe.

8. Darwin's Theory of Evolution by Natural Selection

It is science, not faith, that has forced the abandonment of the pre-Darwinian view that the world is young, and that living organisms were created in their present form by intelligent design.

In the commentary section [6] of the Santa Barbara News-Press, it was claimed that Darwinism (evolution) has its basis in faith, not in science.

On the contrary, evidence for evolution is everywhere. Hiking over mountains, tramping through fields or strolling in gardens, diversity in plants and animals is evident.

In the mesa cliffs above the beach, we see the fossils of sea life in the strata of soils laid down in ancient times. Histories of past events are displayed in cuts made for highways through the surrounding hills and mountains.

Ages of the deposits have been accurately determined from the naturally occurring radioisotopes of the fossils and the soil.

Fossil bones of animals trapped in the La Brea Tar Pits in Los Angeles can be seen in the pits and others are displayed in the museum. Fossils of dinosaurs, plants, insects, mammals, fish, crustaceans--the list is endless--are displayed in museums around the world. All are evidence of evolution.

Darwin's theory of evolution states that organisms in nature evolve by natural selection. This selection depends upon the variability in organisms which significantly enhances the possessor's likelihood of survival and reproduction.

Variations are provided by mutations of the genes. They occur by chance. Most mutations are detrimental and the organisms die, but occasionally the mutations are advantageous and are passed on to the next generation.

In bacteria, maize and man, the rates of mutation are about the same, around two millionths of a mutation per gene pair per generation per individual--sufficient to give the required variations. But natural selection controls the speed, direction and intensity of evolution. Species survive and reproduce because of the benefits of the particular variations and not by chance.

For Darwin's theory of evolution by natural selection, or any other theory of evolution, to be acceptable in science--it must be testable. It has to be falsifiable--that is, capable of being proven false.

A scientific theory makes predictions or explains events in such a way that experiments can be performed or observations carried out that verify or deny its correctness. Before a theory is accepted it is tested in a variety of ways by a number of scientists. If it fails any test it will be modified or discarded.

It is estimated that over 30 million species inhabit the earth today. Many times this number have lived in the past and are now extinct. Over 200,000 species have been described.

The age of the earth is about 4.5 billion years. It appears that life began as early as 3.8 billion years ago, as soon as oxygen was available in the earth's atmosphere.

Bacteria and blue-green algae apparently flourished 2 billion years ago. Many animal species were present 600 million years ago. But humans evolved relatively recently, about 2 million years ago.

Thousands of man years have been spent on tests of the theory of evolution by natural selection. Among the scientific disciplines that have carried out experiments or made observations, who have taken the data and obtained the facts, are Biochemistry, Biogeography, Comparative Anatomy, Embryology, Ethnology, Paleontology and Serology.

Molecular Biology, using DNA identification, has recently verified and discovered new evidence for evolution by natural selection and will make additional contributions in the future.

Complaints are sometimes heard about the lack of evidence for vertical evolution in the geological record. But there are abundant fossil records in the animal, plant, fungus and protista kingdoms (the largest groupings of organisms) over times of hundreds of millions of years.

And there is persuasive evidence that humans and their relatives branched away from chimpanzees about 6 million years ago. The oldest known hominid (human ancestor) fossil, until recently, was Australopithecus afarensis (Lucy) found in Ethiopia that lived about 3.6 million years ago.

Tim White and his colleagues [7] discovered 17 hominoid fossils at Aramis, Ethiopia that were estimated by radioactive dating to be 4.4 million years old. They have named the species A. ramidus and suggest that it is the closest species yet to the split between the human and chimpanzee lines.

Until about 1850 the British peppered moth was observed only with a grey color which could blend in with the trees and escape detection by birds. But with the industrial revolution, the soot killed the lichen on the trees and blackened the trunks and the branches. A black mutant of the peppered moth was favored, and by 1950 all peppered moths were black. When times for each generation are short, such evolution by natural selection, in action, has been seen in over 100 species of insects and spiders.

9. The Dinosaurs' Demise

The earth was submerged for decades in freezing temperatures. Plants and animals died. Half the species on earth became extinct because the organisms couldn't adapt to the harsh conditions in such a short time. The dinosaurs apparently starved and froze to death.

The fossil remains of Gigantosaurus carolinii, likely the largest meat-eating dinosaur ever, were recently discovered by an amateur fossil hunter in the Neuquen province of western Argentina.

The dinosaur was 41 to 43 feet long, weighed 6 to 8 tons, ran on 2 legs and lived 100 million years ago. It is considered bigger than but not related to Tyrannosaurus rex, the former record holder.

During the Mesozoic era, about 248 to 65 million years ago, the land mass Pangaea separated by continental drift into individual continents with their distinctive flora and fauna. This 'Age of Reptiles' is divided roughly equally into Triassic, Jurassic and Cretaceous periods.

Dinosaurs, Greek for 'terrible lizards', are reptiles that were the dominant land animals during the Mesozoic era. They belong to the related orders, reptilelike Saurischia and birdlike Ornithischia. Plesiosaurs and Ichthyosaurs dominated the oceans.

Although dinosaurs are extinct today, their lineage has survived as crocodiles, alligators and birds--sometimes called 'feathered dinosaurs'.

The earliest fossils of dinosaurs appear in the late Triassic period in red-rock deposits in Europe, North America, western China and South Africa. Remains are rare from about 210 to 150 million years ago but then reappear later in the western United States and Tanzania.

Fossils from dinosaurs that lived before about 65 million years ago are widespread over the rocky mountain states in America, Alberta in Canada and Mongolia in Asia. Then they suddenly disappeared. It was the end of the age of dinosaurs and the beginning of mammal development and diversification.

Geologist Walter Alvarez, his father--Nobel prize winning physicist Louis Alvarez--and coworkers in 1980 suggested that a giant asteroid which collided with the earth caused the demise of the dinosaurs.

They discovered a layer of iridium enriched sediment formed 65 million years ago at several sites and suggested that the source of the iridium was an asteroid that had collided with the earth. Independent confirmation came from microspherules, soot and shocked quartz in the vicinity of the iridium.

Soil deposits are built up in time by erosion with the oldest deposits at the deepest levels. Studies in the Hell Creek Formation of Montana and North Dakota showed abrupt extinction in the depth distribution of dinosaur bones.

Above the iridium layer, at times less than 65 million years ago, no bones were found. At the deeper levels the diversity of bones was in agreement with the expected time periods of dinosaur evolution.

But where was the crater that resulted from the asteroid collision?

In 1990, geologists reported an ejecta layer, 0.5 meter thick in Haiti, east of Cuba. They concluded that the collision produced a 300 kilometer crater in the northeast Colombia Basin.

Others found a 350 meter thick ejecta bed in Cuba. They pinpointed the impact crater to the south of the western end of Cuba.

The latest evidence using lead-uranium isotope ratios supports a location identified by a buried impact crater at Chicxulub on the Yucatan Peninsula of Mexico.

The asteroid was estimated to have a diameter of 20 kilometers, a mass of one trillion tons and released an energy equivalent of about one billion megaton bombs.

It appears that the asteroid collision vaporized more than 100 billion tons of sulfur from the Chicxulub rocks. Clouds of sulfuric acid blocked the sunlight. Heat reaching the earth was reduced by about 20 percent and the surface of the earth cooled by 20 to 30 degrees.

10. Molecular Adam and Mitochondrial Eve

In the future, rapid progress in molecular evolutionary anthropology is expected to continue to improve our understanding of human evolution.

Recent studies of molecular biology using DNA have clarified the origins of modern humans. Especially useful are the Y chromosomes transmitted from fathers to sons and mitochondrial DNA (mtDNA) transmitted from mothers to daughters.

The DNA in each cell is encoded with a string of molecules much like letters of the alphabet. These make up the genes that carry the evolutionary history of the species. Evolution proceeds by gradual substitution of one molecule for another in the DNA.

Many mutations are injurious and are eliminated by natural selection. Others are beneficial and are kept for better adaptation. Some are adaptively neutral so have little effect on the organism. These are fixed in the species at rates that are approximately constant.

As a consequence, the number of differences between two species is proportional to the time since their divergence from a common ancestor. Each difference is like the tick of a molecular clock.

Francisco J. Ayala, Professor of Biological Sciences at UC Irvine, in the journal Science [8], reviews the use of the molecular clock to determine the origins of humans.

Robert L. Dorit, biologist from Yale, and coworkers from Harvard and the University of Chicago, compared sections of the Y chromosome from 38 men living in widely different geographic regions. They reported in the journal

Nature [9] that the mens' last common ancestor, "Molecular Adam," lived 270,000 years ago.

More recently, also in Nature [10], Michael Hammer, a scientist at the University of Arizona found that Molecular Adam lived about 190,000 years ago and L. Simon Whitfield and co-workers, geneticists at Cambridge University, England, found [11] 37,000 to 49,000 years ago. These values all have large uncertainties so are not considered in disagreement.

A small fraction of the DNA, about 1 part in 400,000, lies in the mitochondria outside the nucleus of the cell. The mitochondria furnishes the energy for the cell. The late Allan C. Wilson, biochemist from UC Berkeley, Rebecca L. Cann, his student and now Associate Professor of Genetics and Molecular Biology at the University of Hawaii and other students of his, pioneered much of the mtDNA research.

Studies [12] of more than 100 geographically and ethnically diverse women show that they are descendents of "Mitochondrial Eve" who lived about 200,000 years ago.

The lineages to Molecular Adam and mitochondrial Eve, above, are "gene genealogies." That is, particular genes are common to their descendants. However, the "individual genealogies" of each of the contemporary men and women include many other ancestors that have furnished each person thousands of other genes.

It is thought that the origin of modern humans started with the divergence of the hominids from chimpanzees about 6 million years ago in Africa. The earliest hominid, Ardipithecus ramidus, appeared about 1.6 million years later. Evolution continued to H. erectus, who originated in Africa, then spread to Indonesia, China, the Middle East and Europe from about 1.8 to 1.6 million years before now. The transition to H. sapiens took place about 400,000 years ago.

Some confusion exists about the Neanderthals who apparently lived in Europe from about 200,000 to 30,000 years ago. They overlap modern humans that appeared at least 100,000 years before. It's not clear whether either form displaced the other by migration or whether they coexisted.

Controversy rules over the origin of modern humans. Proponents of the "multiregional model" contend that migrations and interbreeding between populations caused development of modern humans simultaneously in widely diverse geographical locations.

The "single region model" argues that modern humans first emerged from Africa or the Middle East more than 100,000 years ago and spread to the rest of the world. The evidence seems to favor this view.

The mtDNA, Y chromosome, and other gene data seem consistent with estimates that the total population of hominids has been reasonably constant at about 100,000 individuals for millions of years.

11. Genetic Cloning

The 21st century will bring fundamental changes in animal husbandry. Identical animals have been cloned from colonies of cells grown in dishes in the lab. Banks of sperm stem cells could be valuable for conserving endangered species, preserving characteristics of prize animals, protecting research animals and maintaining the reproductive ability of males temporarily impotent because of injury, disease or medical treatments.

Breeding animals to improve qualities desired by humans dates back to the beginning of civilization, about 10,000 BC.

Dogs were domesticated and bred as companions, hunters, herders and guards; cows for work, milk and meat; horses for work and racing and sheep for wool and meat.

Planned matings were started in Great Britain in the 18th century. Animals with superior qualities were selectively bred to enhance chosen characteristics. One stud could service many mares and one bull many cows.

The scientific understanding of genetics--the study of animal and plant genes and chromosomes--started in the middle 19th century with the work of Gregor Mendel, a monk in Brunn, Austria. He experimented with peas in his monastery garden and developed laws for explaining inheritance.

Inbreeding (mating among close relatives) was initiated to consolidate special traits. When overdone this leads to reduction in fertility and vitality. Cross breeding within species is then necessary to increase the gene pool and strengthen the breed.

Successful artificial insemination of horses, cattle, sheep and pigs was introduced in Russia during the early 1900s. Semen was collected from chosen males, preserved for several days, then artificially injected into many females.

By 1950 semen could be preserved for long times by solid carbon dioxide at -79 degrees centigrade (C) and later by liquid nitrogen lower yet at-196 C. One degree centigrade is equal to 1.8 degrees Fahrenheit. By this method over 10,000 calves per year have been fathered by one bull. And high quality calves have been produced with semen over 10 years old.

Selective breeding is enhanced by polyovulation--production of many ova (eggs)--from superior females. After fertilization by artificial insemination, the ova are collected from the females and transplanted into a large number of ordinary females. This propagates the desired characteristics of the superior ones. Tens of offspring per year with the desired traits of the female instead of the usual one or two, can be produced by this method.

Today it is possible to procreate identical superior animals by genetic cloning. Scientists have broken embryos into pieces, implanted them in separate wombs and produced several identical progeny.

A glimpse of future genetic cloning was reported [13] in Nature by Keith Campbell and three colleagues at the Roslin Institute for agricultural research on animals located near Edinburgh, Scotland.

The scientists separated cells from sheep embryos and grew colonies of identical cells in lab dishes. These were split several times to produce thousands of cells.

Unfertilized eggs were taken from ewes and all chromosomal material removed. The eggs were then fused to the cells and implanted in additional ewes. New embryos grew and were born as identical lambs.

So far the success rate--live births divided by the number of modified embryos--has been disappointingly low. Only five lambs have been born from this procedure. It's not yet clear whether the failures were caused by the repeated manipulations, or by imperfections in the initial cell line.

Another example of future significance, the preservation of sperm stem cells (cells that produce sperm), was reported [14] by Veterinarian Ralph L. Brinster and co-workers from the University of Pennsylvania. These cells are easily preserved by freezing and can reproduce themselves for indefinite storage.

It may even be possible for the sperm cells to be grown in the testes of animals of other species.

12. WAS ANCIENT LIFE FOUND FROM MARS?

The public, Congress and the President seem agreed that obtaining compelling evidence of ancient life on Mars should receive high priority. It now would be prudent for NASA to speed up plans to recover rock samples by robots on space probes to Mars. Even this does not guarantee that sites chosen will provide evidence of past life on Mars.

My much anticipated copy of Science with the extraordinary report [15] of past life on Mars arrived this week.

Publication of the research article, "Search for Past Life on Mars: Possible Relic Biogenic Activity in the Martian Meteorite ALH84001," by Principal Investigator David S. McKay and eight coworkers supported by NASA--followed the earlier NASA press conference by 10 days. The press conference was forced by rumors and an earlier unauthorized release in the publication Space News.

The authors finished their paper with the bold pronouncement, "Although there are alternative explanations for each of these phenomena taken individually, when they are considered collectively, particularly in view of their spatial association, we conclude that they are evidence for primitive life on early Mars."

The evidence cited by the authors for ancient life on Mars includes:

1. Interior fracture cracks in a Mars rock that were sites of secondary mineral formation and possible biogenic (bacterial) activity.
2. Carbonate globules younger than the rock.
3. Images from high resolution electron microscopes of carbonate globules (about 100 nanometers long) that resembled earth microorganisms, biogenic carbonate structures or microfossils. One nanometer is one billionth of a meter and a meter is about 40 inches.
4. Magnetite and iron sulfide particles--one to 100 nanometers in diameter--that could have been the products of reactions known to occur in terrestrial microbial systems.

5. Polycyclic aromatic hydrocarbons--building blocks of life--formed about 3.6 billion years ago along interior fracture cracks associated with the surfaces containing carbonate globules.

McKay and co-authors did not argue the validity that meteorite ALH84001 originated on Mars. Rather that was done earlier in 1994 by meteoriticist David W. Mittlefehldt [16].

The meteorite was found in Antarctica 12 years ago by Roberta Score, a geologist with the National Science Foundation Antarctica program. It is shaped like a potato and weighs about 4 pounds. It is the oldest of 12 meteorites identified as coming from Mars. These account for about 0.1 percent of all meteorite finds.

The identification as Martian is based on the nearly perfect match between the isotopes of gases trapped in the meteorite and the composition of the Martian atmosphere determined by the analyses on the Viking Landers in 1976. From radioactive isotopes, the rock was determined to have crystallized about 4.5 billion years ago. It is the only Mars meteorite found to date with evidence of ancient Martian microorganisms.

From the products of cosmic ray bombardment in interplanetary space, meteoriticists further deduced that the rock was blasted from Mars about 15 million years ago. The meteorite orbited the sun until landing in Antarctica about 13,000 years ago. This age was determined by radioactive isotopes in the solidified skin of the meteorite that was melted by its passage through the earth's atmosphere.

The most persuasive evidence comes from the electron microscope images of possible microfossils 20 to 100 nanometers in length. However, these are 100 times smaller than the smallest accepted microfossils of ancient bacteria found on earth.

There is also the possibility that the globules and magnetites were formed on Mars by chemical--not biogenic systems.

The evidence that ancient life increased with depth in the fissures of the rock seemed to rule out contamination by earth microbes during the 13,000 years it lay exposed to the weather in Antarctica. The authors also took utmost precautions to protect the rock from contamination during handling and experiments in the lab.

It is imperative that analyses of meteorite ALH84001 be repeated by other scientists and other Mars' rocks be tested again for ancient life. Extraordinary claims must be supported by extraordinary evidence.

13. Agricultural Sciences

The future of agriculture will likely be determined by advances in the fields of biochemistry, plant and animal genetics and molecular biology. Applications in biotechnology will inevitably revolutionize agriculture in the 21st century.

Most of us take for granted our plentiful supply of clean, varied, healthful food.

We shop at the supermarket with its endless supply of fruits, vegetables, meats, bread, desserts and other food and household products.

The counters are covered with unblemished fruit and vegetables, cases are filled with choice cuts of meat and fish, and row after row of attractive displays of all kinds of food tempt the customer.

This didn't happen by accident.

Several segments of society have contributed, but the agricultural sciences are largely responsible for the variety, quality and quantity of food available to the public today.

Cultivation of plants and domestication of animals began about 10,000 years ago. Prior to 1800 over half the population had to farm to supply the world's food. Now in America, one farmer produces enough food for 100 people.

Agricultural education and research in the United States began in 1862 when President Lincoln signed the Morrill Act. Thirty thousand acres were granted to each state for each representative and senator, "for the endowment, support and maintenance of at least one college where the leading object shall be......agriculture and mechanic arts."

The Hatch Act followed in 1887. It established research support for the colleges through the U.S. Department of Agriculture. Agricultural experiment stations were set up in all states. The agricultural extension service was established by the Smith-Lever Act in 1914. The Smith-Hughes Vocational

Education Act in 1917 and the Vocational Education Act of 1963 extended agricultural education to vocational, community and four-year colleges.

Plant sciences have improved the quality and quantity of farm products through research on plant breeding, ecology, genetics, nutrition, pathology and physiology. Double cross-breeding of corn has tripled the yields. Type A is cross bred with B to produce AB; C with D to form CD. Then the hybrid AB is crossed with CD to make the super hybrid ABCD.

Soil and water sciences have devised methods to increase soil fertility and conserve soil moisture. Through chemical fertilization crop yields were increased from 50 to 150 percent.

Pest and disease control in plants and animals has been a major problem with farming from ancient times. (Recall the severe locust attacks in Egypt in the 13th century BC and the famine in Ireland caused by the potato blight in 1845.) Many insecticides, herbicides and fungicides have been developed since 1945 to control pests. At first these were considered a panacea for agriculture, but now many are considered detrimental to the environment.

Animal sciences, through animal physiology, breeding, ecology, genetics and nutrition, have found hundreds of nutrients that are beneficial additives to feed for cattle, chickens, hogs, turkeys and fish in farm ponds. These nutrients include amino acids, carbohydrates, lipids, minerals and vitamins.

Selective breeding through the years has developed superior animals to meet human needs. For centuries, long before the science of genetics, horses were bred for strength and size to work in the fields, and for speed and endurance to race at the tracks.

Recently, emphasis has been on breeding hogs and beef cattle for lean meat, hens that lay an egg a day and dairy cows that give higher butterfat and greater quantities of milk on less feed.

Agricultural engineering has been important for increased efficiency in raising more food and designing better shelters for increased numbers of animals. Larger, faster tractors, trucks and combines have enabled fewer farmers to raise more food on larger farms.

Genetic engineers are trying to increase the photosynthetic efficiency of plants, improve their nutritional value, enhance their ability to fix nitrogen as has already been done in legumes, and better resist disease. The engineers hope their research with animals will lead to multiple clones from embryos and monoclonal antibodies for vaccines.

14. The Rad Lab at UCB

It was 1946, World War II was over and UC Berkeley was my choice of Graduate Schools. It had the reputation for the best high energy physics in the world. It had the Radiation Laboratory directed by Nobel Laureate Ernest O. Lawrence with the new 184 inch proton cyclotron and a strong physics faculty.

During the war as a young Lt. j.g. in the Navy, I had been assigned to the Naval Research Lab and sent to the Philadelphia Navy Yard to work with a small group of scientists under the direction of P. H. Abelson on the separation of uranium-235 from uranium-238 by thermal diffusion. This was the only Navy program in the Manhattan Project for the development of the atomic bomb.

I had applied for assistantships at Berkeley's Physics Department and the Rad Lab, but was rejected by both, each thinking the other had made me an offer. The confusion was cleared up by Raymond T. Birge, Chair of the Physics Department, well liked by the students, who started his letter with, "There's been a comedy of errors in your case." Then offered me a teaching assistantship in the Physics Department.

Lawrence, who received the Nobel Prize in 1939 for invention of the cyclotron, had the reputation as this century's greatest physics entrepreneur. After his first cyclotron in 1932, he built progressively larger and larger cyclotrons with private and public funding. He was the founder of 'Big Science.' His ingenuity was legend. During the war he salvaged magnets from his cyclotrons to use as his 'cauldrons' in the Manhattan Project's uranium isotope separation program. When electrical engineers were scarce, he recruited from Hollywood movie staffs.

He was always one jump ahead of the rest of the field. He made promises in Washington D.C. most scientists thought impossible, then returned to Berkeley and in a few weeks or months delivered. The accelerators at the Rad Lab ran 24 hours a day, 365 days a year. Graduate students and scientists worked all hours of the day and night.

Lawrence made the rounds of the lab at least once a week, sometimes at midnight, asking what they were doing. One Christmas Eve he dropped by the proton synchrotron during its initial tests and was shocked when no one was there. He liked to sit at the controls of the cauldrons and turn the dials full on. Some staff called him 'clockwise Lawrence'.

Birge had recruited Lawrence to Berkeley in 1929 and together they attracted bright young faculty, like Ed McMillan in 1932 and Louis Alvarez in 1936, both later became Nobel Laureates for their research at the Rad Lab.

Many considered Alvarez the best experimental physicist of his time. He was quite supportive of the good graduate students but quick to criticize loose or sloppy work and talks. Even the wise and famous sometimes got 'put down'. Felix Bloch, Nobel Laureate from Stanford, was giving a colloquium at Berkeley. Alvarez was giving him a bad time with questions from the front row. Finally Bloch cut him off with "Oh Louis, you know better than that."

J. Robert Oppenheimer, Director of the Manhattan Project at Los Alamos, N. M., was on the campus my first year before he left to head up the Institute for Advanced Study at Princeton, N. J. As a new graduate student I was awed by his brilliance. Often, after a lecture in the physics department colloquium, much to the speaker's discomfort, Oppenheimer would distill and clarify the entire talk with a concise five minute summary.

After 2 years as a teaching assistant, I moved to the Rad Lab and started research on pi mesons with experiments at the 184 inch cyclotron. The pi meson had been discovered only a short time before, identified in nuclear emulsions exposed to cosmic rays high in the Alps. Using similar techniques they were then found from bombardments by protons at the 184 inch cyclotron.

The Rad Lab was an exciting place for graduate students. It seemed new discoveries were made daily. Weekly Colloquia in the physics department, Journal Club on Tuesday nights and research talks at the Rad Lab were lively and stimulating. Famous visitors from the United States and abroad streamed through the Lab.

In 1948, the 300 MeV electron synchrotron came on line using the synchrotron principle discovered by McMillan. The 300 MeV electrons produced high energy gamma rays. I used these gamma rays to bombard deuterium and study pi mesons for my dissertation under the direction of McMillan, and obtained my Ph. D. in the Spring of 1951.

15. The 1995 Nobel Prizes in the Sciences

The 1995 Nobel Prizes have just been awarded and Scientists in California again lead the World.

Six out of eight of the winners are American, four are professors at universities in California and three are located in the Los Angeles area. UC Irvine claims its first two Nobel Laureates, one in physics and the other in chemistry.

Fred Reines at UC Irvine, and Martin Perl at Stanford split the Nobel Prize in Physics; F. Sherwood Rowland at UC Irvine and Mario Molina at the Massachusetts Institute of Technology share the prize in chemistry with Paul Crutzen at the Max Planck Institute of Chemistry in the Netherlands.

Edward B. Lewis at the California Institute of Technology and Eric Wieschaus at Princeton University share the prize in physiology or medicine with Christiane Nuesslein-Volhard at Tubingen, Germany.

The Nobel prizes are the most prestigious and highly regarded of all international awards. They were established in 1901 by the heirs of Alfred Nobel, a Swedish chemist, engineer and industrialist who invented dynamite.

The prizes are awarded by the Royal Swedish Academy of Sciences from nominations by scientists throughout the world. The financial award this year, about $1 million for each of the 3 disciplines, is divided equally among the winners.

Reines was co-discoverer of the electron neutrino in 1956. It had been predicted in 1931 to balance energy when an electron was emitted from the nucleus of an atom. Later, 2 other kinds of neutrinos were added, the muon and tau neutrinos.

Neutrinos are very difficult to detect. They have no mass and no charge. Their interaction with matter is so low they can escape from the center of the sun and pass through the earth without a collision. Millions of neutrinos from the sun pass through our bodies every second without interacting.

To detect the neutrino, Reines placed a large liquid scintillator close to the Savannah River nuclear reactors in an extremely high neutrino flux. Occasionally, the correct signature appeared in the scintillator to signal the interaction of a neutrino.

Fred Reines and his team also detected the first extragalactic neutrinos, a burst from the explosion of Supernova 1987A that occurred in the Large Magellonic Cloud. He made many other contributions to neutrino and muon physics. His award was long overdue.

Perl discovered the "tau particle," a member of the lepton family, which is similar to an electron but 3,500 times heavier. Prior to his discovery in 1974, the leptons included only the electron, muon, and their 2 neutrinos.

He used the Stanford Positron-Electron Asymmetry Ring and the Electron Linear Accelerator to search for additional leptons. Perl identified 24 particles with the expected properties of the tau lepton. The standard model of elementary particle physics then required an additional tau neutrino for a total of 6 leptons, still the number today.

The Nobel chemistry committee cited Rowland, his former student Molina and Crutzen for "their work in atmospheric chemistry, particularly concerning the formation and decomposition of ozone" that helped prevent "a global environmental problem that could have catastrophic consequences."

Ozone is a very active compound of 3 oxygen atoms, one more than the molecular oxygen in the atmosphere we breath. In the stratosphere, about 50 kilometers (30 miles) up, the ozone layer absorbs much of the ultraviolet radiation from the sun.

In the 1970s these chemists deduced that escaping chlorofluorocarbons, CFCs, like freon from refrigerators, rise and interact with the ultraviolet radiation. Chlorine is released that reacts with the ozone, depleting the ozone layer. CFCs have since been banned for use as refrigerants.

Lewis carried out pioneer work that showed genes control the development of organs in insects and animals. Wieschaus and Nuesslein-Volhard followed with important contributions. All 3 made their discoveries with experiments on fruit flies.

With public demand and support for excellence in science education, this dominance in the sciences could continue well into the next century.

16. SPACE TRAVEL

Don't hold your breath until we visit the stars.

The ultimate dream of the space enthusiast is to visit another intelligent civilization. Astronomers have been scanning the skies with optical telescopes and listening intently with radio telescopes, but no civilization has turned up yet.

However, the lack of contact shouldn't dampen our fervor or stifle our fascination. Yet before contacting our local travel agency, it may be prudent to inquire about the difficulties and the costs of the flight.

Most remember the thrill of the first moon-landing on July 20, 1969 with the historic message from Neil Armstrong that "the Eagle has landed." Five successful flights followed and few begrudge the tens of billions of dollars spent on these missions.

For scientific exploration of the planets, NASA wisely sent unmanned spacecraft only. But a manned flight to Mars, if desired, may be possible in the 21st century. The distance to the Moon is about 400,000 kilometers (1.6 kilometer equals 1 mile) and to Mars about 120 million kilometers at nearest approach to the Earth.

If the spacecraft to Mars travels at the speed of Apollo to the moon it would take about 3 years, one way. To protect the travelers from the space environment and to furnish supplies to sustain life, a much larger, heavier, sophisticated spacecraft is required.

In addition, the booster for the return trip to the earth would need to be much bigger and heavier than the one from the moon because of Mars' much larger gravitational attraction. A very large initial boost from the earth would be needed. Although the voyage should be technically feasible, the costs could be 100s of billions of dollars.

Trips to the stars are not now and may never be feasible. They are limited not only by technology but by fundamental laws of physics that cannot be violated. Information, which includes spacecraft, cannot be transmitted faster than the velocity of light, 300,000 kilometers per second.

A consequence of Einstein's Special Theory of Relativity is that an event in the frame of reference of a body moving at a speed close to velocity of light, suffers a time dilation. Events take longer than in the frame at rest. The heart beats slower for the person in the frame of the spacecraft than in the frame of the person that stays on the earth. So the astronaut on the space voyage returns younger than his twin who stayed home. This is an aid to the traveler, as he can make longer trips in his lifetime than he otherwise could.

Likewise, the spacecraft and payload increase in mass over the mass in the rest frame as they approach the speed of light. In attempting to increase the spacecraft speed, much of the energy goes into increased mass instead. As a consequence, for longer and longer trips when the speed of the spacecraft is increased to enable the astronaut to make the trip in his lifetime, the energy required eventually becomes impossibly large.

Suppose a trip is planned to a planet (not yet discovered) of the nearest star, Alpha Centauri. It is four light years away. One light year is the distance light travels in one year. Alpha Centauri is 100 million times farther away than the moon. At the same speed as the Apollo spacecraft, the journey would take one million years.

Let's try a speed of one-tenth the velocity of light. Then the flight time is 40 years in the earth's frame and slightly less, 39.8 yr., in the astronaut's frame. The relativistic time dilation is small but the energy expended is immense.

Assign a ridiculously low mass to the spacecraft with its passengers, supplies, and protection from the space environment of 10 tons, equivalent to 10 small cars. No booster fuel is included. The added mass energy is five tenths of one percent. That is equal to two percent of the total electrical energy generated in the United States in 1993. And this estimate did not include the energy necessary to slow the spacecraft for landing on the planet nor for the return flight to the earth.

It is assumed that the boost to one-tenth the speed of light would come from radio waves focused on a spacecraft antenna from radio stations on the earth or by some other technology, neither yet invented.

And this trip may not seem rewarding to the astronaut as he would not be able to spend time on the planet and return within his lifetime.

Many other possible flight plans could be considered. Suppose the spacecraft speed were increased instead to 99 percent of the velocity of light. Then the trip to the planet would take only six-tenths of a year in the astronaut's frame. He could visit the planet and return in a small fraction of his lifetime. He would be 6.8 years younger than his twin on return.

But the penalty paid in boost energy would be prohibitive. With the same conditions as the flight at one-tenth the speed of light, an energy equal to 20 times the total electrical energy produced in the United States in 1993 is required.

CHAPTER VII

SCIENCE EDUCATION

In the preceding six chapters, reasons have been given why science has been important in our lives and why science is needed to solve current and future personal and public problems. I hope these arguments have been persuasive and readers are convinced of the need for increased education in science and math at all levels. But before addressing the issue of science education we review the public's current state of knowledge and its appreciation of science.

THE PUBLIC'S VIEWS

The public continues to hold science in respect, sometimes even in awe. According to the National Science foundation "Science and Engineering Indicators 1996" [1], three-quarters of the U.S. population believe that the benefits of science research outweigh its harmful results with only about 15% holding the opposite view. These opinions have been nearly constant since the first survey in 1979. There seems to be little evidence of an anti-science movement in the findings.

About 45 percent of the respondents [2] were "very interested" in scientific discoveries and an additional 45 percent "moderately interested." Fifty-three percent expressed strong interest in the environment and 41 percent moderate interest.

Interest and awareness were closely proportional to level of the respondent's education--the higher the education the greater the interest. The study found that the public gets most of its science information from newspapers and TV. Only about 10% come from science publications.

Scientists and science leadership are highly regarded. Thirty-eight percent said they were "very confident" of the leadership of the scientific community. The rating was significantly higher than that of the U.S. Supreme Court, the military, educators and leaders of organized religion. Only 8 percent expressed confidence in journalists and members of Congress.

Jan Miller [1] finds that, "more than eight of ten Americans believe that science and technology continue to make their lives healthier, easier, and more comfortable, reflecting nearly two decades of positive assessment of the net impact of science and technology on their lives and society."

However, this positive feeling of the public toward science must be qualified by its limited understanding of science. According to Miller, "Americans have a huge amount of confidence in scientists. But the basis of that support is faith, not knowledge...... We are a democracy, but only 9 percent of the population can participate in a serious science or science policy debate." A much higher level of public understanding of science is required to set policy on the large number of issues that confront the electorate--many of which involve science.

The public has little comprehension of how scientists carry out experiments, make observations, come to conclusions and arrive at theories. It is difficult for the non-participant to imagine the numerous checks and balances built into scientific procedures in an attempt to validate results and avoid errors. This occurs before publication of the results and distribution of information to the public and continues after publication until the evidence is overwhelming. When there are disagreements among scientists, often occurring at the frontiers of research, it is difficult even for scientists to decide who is right. But in most other situations where division occurs, informed citizens should be able to evaluate the evidence and decide on the proper course of action. Here, a better background of science literacy, skills and methods of evaluation are needed.

EDUCATION

How do we improve the public's understanding of science? It's simple to describe but difficult to accomplish. It is through "Education, education and more education." More and better science education at all levels from elementary grades through intermediate and high schools and community and four-year colleges. And lifetime learning for adults. When our lives and welfare depend on science, that's not too much to ask. It means more hours in science classrooms, more dedicated teachers with science training and more encouragement and support from parents in the home.

While our primary concern here is science education for all students, more and more jobs these days require science preparation. The Bayer Corporation survey [2], conducted in April 1996 found that 60 percent of corporate human resource directors and 40 percent of elementary school principals said that most students finishing school lack adequate science preparation for entry-level jobs in industry.

Shirley Malcolm, head of the Directorate of Education and Human Resources Programs at the American Association for the Advancement of Science says [2] , "Most people do not get a basic education in the sciences, and they're not getting enough subsequent education and exposure to science."

The National Science Education Standards [3], published early in 1996, give comprehensive guidelines to improve science education from kindergarten through high school. The 262-page Standards "spell out a vision of science education that will make scientific literacy for all a reality in the 21st century.....They point toward a destination and provide a road map for how to get there." Rote memorization will be replaced by interactive learning. "Cookbook" experiments are superseded by self-directed inquiry and problem solving. "Hands-on" and "minds-on" activities will involve all students.

Richard Klausner, chair of the Standards project points out [2], "The National Science Education Standards are designed to stimulate the sweeping improvements necessary to achieve scientific literacy for all students."

Robert Hazen, co-author of the book [2], "Science Matters", emphasizes, "Science is a way of asking questions and answering them,.....a way of understanding our place in the universe and in the world around us. And it is the most exciting ongoing adventure in the human experience."

It follows that new standards for science also require changes in the way science is taught and teachers are trained. "Science and Engineering Indicators 1996" reports that less than 4 percent of elementary math and science teachers majored in math or science, 11 percent of intermediate school math and 21 percent of science teachers majored in their specialized fields. The Standards suggest that effective science teachers need both theoretical and practical training in science and education with lifelong learning. The teachers need to work with other science teachers to use outside resources. They should consider themselves members of the scientific community as well as teachers.

The teachers will need adequate time for preparation, planning and outside activities; resources, materials and facilities for the hands-on approach; and sufficient financial benefits and social status to encourage their whole-hearted participation.

Parent help and advising needs to be improved. According to the Bayer Survey [2], 68 percent of parents do not consider themselves sufficiently knowledgeable to help with their child's science homework. The recommended higher level of science in school and lifelong learning will prepare parents to assist their children with schoolwork as well as participate in public science issues.

Leon Lederman's article in the Skeptical Inquirer [4],"A Strategy for Saving Science," is an incisive analysis of the current health of science and the revolution in science education required to combat parascience, antiscience, TV late night junk science shows, academic reconstructionists and less than adequate science learning by students K-16.

He was a Professor of Physics at Columbia University from 1958 to 1989, Director of Fermi National Accelerator Laboratory at Batavia, Illinois from 1979 to 1989 and currently Pritzker Professor of Science at the Illinois Institute of Technology. He is a member of the National Academy of Sciences and received the Nobel Prize in Physics in 1988. He is currently devoting much of his time and energy to improvement of science education in the schools and lifelong learning by adults, which he calls K-100.

Lederman strongly believes, "To preserve our four-hundred-year commitment to a scientific world view, we need our educated people to incorporate scientific thinking--the blend of curiosity and skepticism, the habit of critical questioning--into their very nature. Pioneering new teaching styles in science and math, carried out in cooperation with the liberal arts, can help achieve that."

Indeed, he states, "the ultimate argument for not abandoning science to the dark forces of superstition, ignorance, and rigid belief systems is that the planet will not survive a population of upwards of ten billion people (by the year 2050?) without significant increases in our knowledge base, without new forms of energy, food production and mechanisms for raising the standard of living of the poorest people."

"We must work together," Lederman says, "--scientists, educators, psychologists, neuroscientists, linguists, anthropologists--to make it better...... The strategic vision is that if an ever-increasing number of our citizens could be taught to think scientifically,.....(they) would be intolerant of sound bites and baloney, would insist on the proper allocation of national resources,....would insist that the products of science and technology be deployed for the long-term benefit of the many, and would understand the role of knowledge in social, economic and cultural contexts."

Lederman sees an education circle that starts with pre-school and passes through elementary school--K-8, high school--9-12, college--13-16 and continues as citizens, parents, teachers, voters and legislators to the end of life--17-100. Because of the central role science and technology play in society, achieving scientific literacy must be central to K-8 education.

High school, 9-12, should have three years each of math and science. But the order of science courses should be reversed from their historical order of biology, chemistry then physics. The current order has been dictated largely by the math capability of the students. But by teaching the necessary math earlier and concentrating on concepts, the new order makes more sense.

The student starts in physics with an understanding of atoms as the basic structure of matter and fundamental forces and laws like gravitation, conservation of energy and electromagnetism. The chemistry of atoms, molecules and compounds and their chemical reactions then follows naturally. The more complicated biology can be understood on the basis of chemistry and evolution. In this order the students are prepared for the marvelous new discoveries of molecular biology. Astronomy, including the solar system, can be taught with physics and the applicable parts of earth sciences in each of the three courses.

Having taught astronomy for several years to humanity students at UC Riverside, I am well aware that many are deficient in math and science. This occurred even though the students were in the upper 12.5 percent of their high school graduating classes. Coming out of a stronger math-science curriculum, these students would be better grounded in math and science and would find material at the level of this book quite easy. I would hope they would find courses in colleges and universities on issues similar to those raised by "Why Science?" and would be challenged to apply science to help solve problems of society. In agreement with Lederman, two years of science in 13-16 is a bare minimum.

This book is directed toward the educated public along with the teachers of science at all levels. It is the citizens who are confronted with difficult issues at the polls and a society that requires a knowledge of science and technology. Many issues could be added to those addressed in columns in this book. A few more raised by Lederman [4] are: Should we ban cigarettes? Can peach pits cure cancer? Should we tax carbon emissions? Are humans influencing global climate change? Should we decriminalize drugs? How do we understand and control the information revolution?

THE MEDIA

1. Television. "In the typical home, television is on an average of seven hours and 41 minutes a day," according to George Gerbner formerly Dean of the Annenberg School for Communication at the University of Pennsylvania [5]. He was the person most responsible for establishing communications as an academic discipline. Students spend almost as much time watching TV as they spend in the classroom. Therefore much of their science is learned from TV. Except for the news, TV networks seem to feel little or no responsibility for accuracy in the science depicted in their programs. After all, the programs are just for entertainment. At best the science is distorted but more often it is pseudo- or anti-science.

After NBC showed the program, "The Mysterious Origins of Man" in February, 1996, scientists objected that it promoted pseudoscience and misled the public. With commentary by Charlton Heston, the program suggested that evolution is a questionable theory, human civilizations began more than 100 million years ago, and scientists conspired to suppress important archaeological information. Of course these were blatantly false. NBC then showed its contempt for the criticism by rerunning the program in June billed as a "program that dares to challenge accepted beliefs." Bill Cote, the program's independent producer said, "NBC's extensive legal department put us through the wringer until we presented a balanced view."

Have you been kidnapped and assaulted by a space alien today? It's as American as say, "you know." And it's a great way to make a living. You can make the rounds of the talk shows and repeat the same preposterous tales at each stop--even get invited back to the "Larry King Live" show several times.

An exception to the shenanigans is public television station KCET. Blaine Baggett, vice president of program development, scheduling and acquisition for KCET, states that KCET is committed to science [6]. He points out that "Nova," the best science series in all of television, appears on KCET 52 weeks a year. Furthermore, KCET is proud that it has produced a few of Nova's award-winning episodes including "smart" weapons used in the Gulf War and the Northridge earthquake.

KCET has a long exemplary record making science documentaries. National releases include "Cosmos" with Carl Sagan in 1980 that set a standard for science television. One of television's highest-rated miniseries was "Discoveries Underwater" in 1988. "The Astronomers" continued the celestial exploration in 1991 with local astronomical expertise. Recently KCET produced Roger Bingham's multi-part series, "Human Quest," involving psychology and the intricate workings of the brain.

Local productions include the Emmy Award-winning three-part miniseries "L.A. Medical," part of KCET's "Life & Times," "Mystery of Pygmy Mammoth,", "Return to Mars," "Undersea Earthquakes," "Richard Feynman's Lost Lecture" and others. Some of the world's most renowned scientists, Donald Johanson, Edward Teller, Francis Crick, Sylvia Earle, Leonard Kleinrock, Carl Sagan and others have shared their views on "Life & Times."

In addition, many science programs produced elsewhere such as "Nature," "The New Explorers," "Future Quest," "21st Century Jet" and "Life on the Internet" are broadcast by KCET. The shows "Newton's Apple," "Let's Go Fly a Helicopter," "The Magic School Bus," "Bill Nye the Science Guy" and "Kratt's Creatures" provide excellent science viewing for children.

2. The Press. A second source of science information available to and used by the public is newspapers. The science appearing in most local papers is picked up from the newspaper wire service. However, local newspaper staff often report local science items. In either case the coverage is not as extensive as in the large newspapers which have their own science writers.

Too often, the science material is reported like other news in the paper. The reporter interviews the person who is the primary subject of the story and writes up what he or she has been told. With little science background it is easy to be fooled, or conned by the subject. The story may make little sense at best and at worst is complete nonsense. Such is the case of a story in the Santa Barbara News Press about astral projection (out of body travel) under a "science" header by Lydia Martin of the Knight-Ridder News Service [7]. The International Institute of Projectology was to be set up in Miami, Florida. The Projectology instructors said you could learn the tricks to "soul tripping" in a few easy lessons. We don't need this space age religion.

The Press has not escaped the current irrational alien kidnapping rage. Can you believe that the Santa Barbara News Press ran a story [8]—straight--about Dr. Roger Leir, a podiatrist from Thousand Oaks, California who removed three metal fragments from two patients. The man and woman believed they had been abducted by space aliens. The article announced that Dr. Leir would lecture on Saturday night about the fragments "that could yield the kind of physical proof that has been lacking. (Tickets are $10 at the box office.)" Dr. Leir states, "We know now that we have some very interesting objects that are probably extraterrestrial."

"You have to form a theory, and then take it into a laboratory and either prove it or disprove it, and that's the stage where we are," Dr. Leir said. He has been lecturing nationwide on his findings. Dr. Leir seems to have put the cart before the horse. The usual scientific procedure is to do the necessary tests in the lab first, give talks to other scientists and at meetings (at no charge) for comments and criticisms, next submit a paper about the discovery to a peer-reviewed journal like Science or Nature for publication. And finally, after publication, issue the press release.

Of course, if Dr. Leir really has the smoking gun, it is the greatest scientific discovery the world has ever known. Instead of giving lectures at

$10 per head (not a bad living if anyone shows up), he would receive the Noble Prize worth $1 million and be wined and dined by kings and heads of state.

Some of the major newspapers in the country like the New York Times and the Los Angeles Times have their own staff science writers. These writers usually receive early press releases about discoveries from the scientists, their home organizations or support agencies. The writers interview the scientists and other experts; then release their stories on the day the research is published. The stories are usually well written, descriptive, accurate and at a level that knowledgeable readers can understand. Although at times, the order of presentation of the material is a bit frustrating. It may be necessary to go to the end of the article to find what the scientists really did that was new and different.

Among the excellent science writers covering research for the New York Times are: Natalie Angier, William J. Broad, Jane Brody, Malcolm Browne, Youssef Ibrahim, Gina Kolata, Warren Leary, John N. Wilford and Carol Kaesuk Yoon. For the Los Angeles Times they are: K. C. Cole, Marla Cone, Robert Lee Hotz, Thomas Maugh III and Robert Rosenblatt.

The science writers are very influential because it's often the only science--except for very limited accurate science on TV--that the public ever sees. Unfortunately, occasionally even one of these writers goes off the deep end. William Broad contributed significantly to the delay of the--still not opened--much needed nuclear waste repository at Yucca Mountain. He publicized and promoted incorrect renegade scientists' claims of ground water problems in a feature story, "A Mountain of Trouble" in the New York Times Magazine in 1990 [9] and of possible nuclear explosions in a front page story in the New York Times headlined "Scientists Fear Atomic Explosion of Buried Waste; Debate by Researchers; Argument Strikes New Blow Against a Proposal for Repository in Nevada," in 1995 [10].

A number of scientific journals (magazines) are available to the reader interested in science. I have referred to many of these in the columns. The "Skeptical Inquirer" is recommended for articles on the paranormal, parascience and antiscience. "Sky and Telescope" is excellent for astronomy and astrophysics. "Physics Today" has readable articles in physics and comes free to members of the American Physical Society. "EOS", in newspaper format, has articles about the earth sciences and is free to members of the

American Geophysical Union. "Scientific American" features review articles for the public by outstanding scientists in their fields.

My favorite for keeping up to date on many fields of science is "Science," published by the American Association for the Advancement of Science, sent free to its members. It is a fast-publication journal and has the highest readership of any scientific journal in the United States. Because of its reputation, Science is often preferred by research scientists for reporting new discoveries and scientific breakthroughs. It features science news and contains brief reviews, in readable style, of some of the contributions. But most of the articles and reports are not for amateurs. "Nature" is the British counterpart of Science. It also has a very large worldwide readership.

There is no question about the positive impact science has had on the lives of people in the past. I am optimistic about its contribution to our lives in the future. With the greater emphasis on math and science education and with a more enlightened public, we should be able to solve many of our private and public problems.

Some world problems seem intractable. Science has not always been used wisely by the public, our leaders--or the scientists, themselves. But the fear-mongers, and activists who are scaring the public with non-existent, exaggerated or incorrect claims involving science, can be exposed by a well-informed society. The nonsense bombarding us constantly from the purveyors of the paranormal, pseudoscience, and antiscience can be countered by lifelong science education. Assisting the public in solving their problems requiring science and providing information for teachers of science is my goal and the intent of this book.

References

Introduction

1. George H. Gallup, Jr. and Frank Newport, Skeptical Inquirer, Vol. 15, No. 2, p.137-146, Winter 1991.

Chapter I

1. Jeremy Wallace in the column "Surveying Generation X", Santa Barbara News Press, September 27, 1994 gives results of a poll released by the advocacy group, Third Executive Director, Richard Thau; poll taken by Frank Luntz and Mark Siegel.
2. Christopher French and co-authors, Skeptical Inquirer, Vol. 15, No. 2, p. 166-172, Winter 1991.
3. Lydia Martin, writer for Knight-Ridder News Service, Santa Barbara News Press, August 15 1995.
4. Leonard Angel and others, Skeptical Inquirer, Vol. 18, No. 5, p. 481-487, Fall 1994; and Vol. 19, No.2, p. 56, March/April 1995.
5. Lloyd Stires and Philip Klass, Skeptical Inquirer, Vol. 17, No. 2, 142-146, Winter 1993.
6. Robert Sheaffer, Skeptical Inquirer, Vol.19, No. 1, p. 21, January/February 1995.
7. William Grey, Skeptical Inquirer, Vol. 18, No. 2, p. 142-149, Winter 1994.
8. International Herald Tribune, April 1991.
9. Russell Targ and Harold Puthoff, Nature, Volume 251, p. 559, October 18, 1974.
10. Rikki Razdan and Alan Kielar, Skeptical Inquirer, Vol. 9, No. 2, p. 147-158, Winter 1984/85.

11. Michael Dennett, Skeptical Inquirer, Vol.13, No. 3, p. 264-272, Spring 1989; and Vol. 18, No. 5, p. 498-508, Fall 1994.

12. P. E. Damon and 20 authors, Nature, Volume 337, page 611-615, February 16, 1989.

CHAPTER II

1. Paul Gross and Norman Levitt, "Higher Superstition," The Johns Hopkins University Press, Baltimore and London, 1994.

2. Bernard Ortiz de Montellano, Skeptical Inquirer, Vol. 16, No. 1, p. 46-50, Fall 1991.

3.Constance Holden, Science, Vol. 271, p. 1357, March 8, 1996.

4. Francis B. Harrold and Raymond A. Eve, Skeptical Inquirer, Vol.11, No.1, p. 61-75, Fall 1986.

5. George H. Gallup, Jr., and Frank Newport, Skeptical Inquirer, Vol.15, No. 2, p. 137-146, Winter 1991.

6. Jeremy Rifkin, "Beyond Beef", Dutton, New York, 1992.

7. Alan Sokal, "Transgressing the Boundaries: Toward a Transformative Hermeneutics of Quantum Gravity," Social Text, p. 217, Spring/Summer 1996.

8. Alan Sokal, Lingua Franca, p. 62, May/June 1996.

9. Stanley Fish, New York Times, May 21, 1996.

10. M. Kreitzer, New York Times, May 23, 1996.

11. Jerry Coyne, New York Times, May 23, 1996.

12. Ruth Rosen, Los Angeles Times, May 23, 1996.

13. Jacque Benveniste and 12 authors, Nature, Vol. 333, p. 816, June 30, 1988.

14. Martin Gardner, Skeptical Inquirer, Vol. 13, No. 2, p. 132-141, Winter 1989.

15. Eric Lander and Bruce Budowle, Nature, Vol. 371, p. 735-738, October 27, 1994.

CHAPTER III

1. For an excellent review of the history of medicine see the 1994 edition of the Encyclopedia Brittanica.

2. Barry Beyerstein and Wallace Sampson record their impressions of current TCM in China in the Skeptical Inquirer, Vol. 20, No. 4, p.18-26, July/August 1996.

3. Eliot Marshal, "Science and Comment", Science, Vol. 265, p. 2000-2002, September 30, 1994.

4. Gina Kolata, science reporter for the New York Times, June 17 and 18, 1996.

5. Bela Scheiber and Carla Selby, Executive Director and Special Projects Director for the CISCOP Center for Inquiry--Rockies, describe TT therapy in the Skeptical Inquirer Vol. 20, No. 4, p.15-17, July/August 1996.

6. The Los Angeles Times, June 16, 1995.

7. Richard Ofshe and Ethan Watters, "Making Monsters: False Memory, Psychotherapy and Sexual Hysteria," Charles Schribners Sons, New York, 1994.

8. Elizabeth Loftus and Katherine Ketcham, "The Myth of Repressed Memory", St. Martins' Press, 1994.

9. Mark Pendergrast, "Victims of Memory: Incest Accusations and Shattered Lives," Upper Access Books, 1994.

10. Martin Gardner, Skeptical Inquirer, Vol. 17, No. 4, p. 370-375, Summer 1993.

11. Philip Rosenberg, Science, Vol. 270, p. 1372-1375, November 24, 1995

12. Jon Cohen, Science, Vol. 266, p. 1645-1646, December 9, 1994.

13. David Ho and co-authors, Nature, Vol. 373, p. 123-126 and Xiping Wei and co-authors, p. 117-122, January 12, 1995.

14. Joseph E. Murray, Los Angeles Times, February 5, 1996.

15. John Collinge and coworkers, Nature, Vol. 378, page 779-783, 21/28 December 1995.

16. "Health Effects of Low-Frequency Electric and Magnetic Fields," prepared by the Oak Ridge Associated Universities Panel for The Committee on Interagency Radiation Research and Policy Coordination, ORAU92/F8, June 1992.

17. William R. Bennett, Jr., "Health and Low Frequency Fields," Yale University Press, New Haven, Connecticut, 1994; and "Cancer and Power Lines, Physics Today, Vol. 479, No. 4, p 23-29, 1994.

18. Glenn T. Seaborg, Skeptical Inquirer, Vol. 19, No. 1, p. 39-40, Jan/Feb. 1995.

19. Peter Huston examined TCM in the Skeptical Inquirer, Vol. 19, No. 5, p. 38-42, September/October 1995. Teams from the Skeptical Inquirer reported about modern TCM in the Skeptical Inquirer, Vol. 12, No. 4, p. 364-375, Summer 1988 and Vol. 20, No. 5, p. 27-34, July/Aug. 1996.

CHAPTER IV

1. David Brower, editorial in the July 22, 1996 issue of the Los Angeles Times.

2. Los Angeles Times diary in the August 1, 1996 issue.

3. Andrew Knoll and three other authors, Science Vol. 273, p. 452-457, July 26, 1996.

4. P. H. Abelson, Editorial in Science, Vol. 260, p.1859, June 25, 1993.

5. Paul Ehrlich, "The Population Bomb," Ballantine Books/Sierra Club, New York, 1968.

6. Paul Ehrlich and Anne Ehrlich, "The Population Explosion," Simon and Schuster, New York, 1990.

7. John Bongaarts, Scientific American, Vol 270, No. 3, p. 36-42, March 1994. Bongaarts has won several awards for his contributions to population control.

8. Bryant Robey, Shea Rutstein and Leo Morris, Scientific American, Vol. 269, No. 6, p. 60-67, December 1993.

9. P. H. Abelson, Editorials in Science: Vol. 259, p. 1235, February 26, 1993; Vol. 265, p. 1507, September 9, 1994; and Vol. 266, p. 1303, November 25, 1994.

10. B. N. Ames and L. S. Gold, Science, Vol. 249, p. 970, 1990; and L. S. Gold et al., Science Vol. 258, p. 261, 1992.

11. Theo Colborn, Dianne Dumanoski and John Peterson, "Our Stolen Future," Dutton, New York, 1996.

12. "Carcinogens and Anticarcinogens in the Human Diet," Committee on Comparative Toxicity of Naturally Occurring Carcinogens, National Research Council, National Academy Press, Washington, D.C., 1996.

13. Ralph D'Agostino, Jr., and Richard Wilson, "Asbestos: The Hazard, the Risk, and Public Policy," in the book "Phantom Risk," edited by Kenneth

Foster, David Bernstein and Peter Huber, The MIT Press, Cambridge, Massachusetts, London, England, p. 183-210, 1994.

14. B. N. Ames and L. S. Gold, Science, p. 970, August 31, 1990.

15. Michael Gough, "Dioxin: Perceptions, Estimates and Measures" from the book, "Phantom Risk," edited by Kenneth Foster, David Bernstein and Peter Huber; MIT Press, Cambridge, Massachusetts, London, England, p. 249-278, 1994.

16. P. H. Abelson, Editorial in Science, Vol. 265, p. 1155, August 26, 1994.

17. Discoverer, Vol. 17, No. 5, p. 82-83, 1996.

18. "Scientific and the Endangered Species Act," Panel of the National Academy of Sciences, chaired by Michael T. Clegg, University of California, Riverside, 1995.

19. Bruce Walsh, Roger Angel and Peter Strittmatter, Nature, Vol 372, p. 215-216, November 17, 1994.

20. "Ward Valley; An Examination of Seven Issues in Earth Sciences and Ecology," Committee to Review Specific Scientific and Technical Safety Issues Related to the Ward Valley, California, Low-Level Radioactive Waste Site, National Research Council, Chair, George A. Thompson, Stanford University, National Academy Press, Washington D.C., 1995.

21. "Ground Water at Yucca Mountain: How High can it Rise?," Final Report of the Panel on Coupled Hydrologic/Tectonic/Hydrothermal Systems at Yucca Mountain, National Research Council, National Academy Press, Washington, D.C., 1992.

22. William Broad, New York Times, March 5, 1995.

Chapter V

1. The American Almanac 1995-1996--Statistical Almanac of the United States 115th Edition, The Reference Press, Austin, Texas.

2. The reports include the 1985 study of the Pennsylvania State Health Department; The report of the Three Mile Island Public Health Fund published in the September 1990 issue of the American Journal of Epidemiology by Columbia University epidemiologist, Maureen Hatch, and coworkers; and the publication in the June, 1991 issue of the American Journal of Public Health by scientists from Columbia University and the National Audubon Society.

3. P. H. Abelson, Editorial in Science, Vol. 272, p. 464, April 26, 1996.

4. Lester Lave and co-authors, Environmental Science and Technology, Vol. 30, No. 9, p. 402A-407A, September 1996.

5. P. H. Abelson, Editorial in Science, Vol. 266, p. 347, October 21, 1994.

6. Los Angeles Times, April 21, 1995.

7. The California Almanac 7th edition, James S. Fay, editor, Pacific Data Resources, 1995.

8. Eric D. Larson "Technology for Electricity and Fuels from Biomass," Annual Review of Energy and Environment, Volume 18, p. 567-630, 1993.

9. John R. Huizenga, "Cold fusion: The Scientific Fiasco of the Century," University of Rochester Press, 1992. Dr. Huizenga is a Professor of Chemistry and Phsics at the University of Rochester. He was Co-Chair of the US Department of Energy—Energy Research Advisory Board Cold Fusion Panel.

10. Los Angeles Times, April 21, 1995.

CHAPTER VI

1. P. H. Abelson, "The Changing Frontiers of Science and Technology," Science, Vol. 273, p. 445-447, July 26, 1996.

2. Eugene Wong, Nature, Vol. 381, p. 187-188, May 16, 1996.

3. U.S. News and World Report, discussed in the Santa Barbara News Press, December 11, 1994.

4. Floyd E. Bloom, editorial in Science, Vol. 270, p. 1901, December 22, 1995.

5. M. H. Anderson and co-authors, Science, Vol. 269, p. 198-201, July 14, 1995; in addition see Gary Taubes, News, p. 152-153 and Keith Burnett, Perspectives, p. 182-183.

6. Santa Barbara News-Press, October 21, 1994.

7. Tim White, Gen Suwa and Berhane Asfaw, Nature, Vol. 371, p. 306-312, September 22, 1994.

8. Francisco J. Ayala, Science, Vol. 270, p. 1930-1936, December 22, 1995.

9. Robert L. Dorit, Hiroshi Akashi and Walter Gilbert, Science, Vol. 268, p. 1183-1185, May 26, 1995.

10. Michael F. Hammer, Nature, Vol. 378, p. 376-378, November 23, 1995.

11. L. S. Whitfield, J. E. Sulston and P. N. Goodfellow, Nature, Vol. 378, p. 379-380, November 23, 1995.

12. Allan C. Wilson and Rebecca L. Cann, Scientific American, p. 68-73, April 1992.

13. Keith Campbell and colleagues, Nature, Vol. 380, p. 64-66, March 7, 1996.

15. David S. McKay and eight co-authors, Science, Vol. 273, p. 924-930, August 16, 1996.

16. David W. Mittlefehldt, Meteoritics, Vol. 29, p. 214-221, 1994.

CHAPTER VII

1. National Science Foundation, "Science and Engineering Indicators 1996". Jon Miller, Vice-President of the Chicago Academy of Sciences, is the main author of the section on the public understanding of science.

2. A review of the "Science and Engineering Indicators 1996" of reference 1 is given by Michael Carlowicz in EOS, Transactions, American Geophysical Union, Vol. 77, No. 35, August 27 and No. 38, September 17, 1996.

3. “The National Science Education Standards,” National Research Council, National Academy Press, Washington DC, 1995.

4. Leon Lederman, Skeptical Inquirer, Vol. 20, No. 6, p. 23-28, November/December 1996.

5. Lee Dye, Science Watch, Los Angeles Times, June 24, 1996.

6. Blaine Baggett, article in the Los Angeles Times, November 4, 1996.

7. Lydia Martin, "Projectology Institute Has Lofty Goals," Santa Barbara News Press, August 15, 1995.

8. Ben Hellwarth, "Space Alien 'Evidence' Offered," Santa Barbara News Press, November 15, 1996.

9. William Broad, "A Mountain of Trouble," New York Times Magazine, November 18, 1990.

10. William Broad, New York Times, March 5, 1995.

INDEX

E

F

G

H

I

J

K

L

M

N

O

P

R

S

T

U

V

W

Z